National Defense Research Institute

ENLISTED MANAGEMENT POLICIES AND PRACTICES

A Review of the Literature

SHEILA NATARAJ KIRBY
SCOTT NAFTEL

Prepared for the
Office of the Secretary of Defense

RAND

The research described in this report was sponsored by the Office of the Secretary of Defense (OSD). The research was conducted in RAND's National Defense Research Institute, a federally funded research and development center supported by the OSD, the Joint Staff, the unified commands, and the defense agencies under Contract DASW01-95-C-0059.

Library of Congress Cataloging-in-Publication Data

Kirby, Sheila Nataraj, 1946–   .
  Enlisted management policies and practices : a review of the literature / Sheila Nataraj Kirby and Scott Naftel.
    p.   cm.
  "Prepared for the Office of the Secretary of Defense by RAND's National Defense Research Institute."
  "MR-913-OSD."
  Includes bibliographical references.
  ISBN 0-8330-2609-7 (alk. paper)
  1. United States—Armed Forces—Personnel management.
  2. United States—Armed Forces—Recruiting, enlistment, etc.
  I. Naftel, Scott, 1952-   . II. United States. Dept of Defense. Office of the Secretary of Defense. III. National Defense Research Institute (U.S.). IV. Title.
  UB323.K56   2001
  355.6 ' 1 ' 0973—dc21   98-3505
                                                                      CIP

RAND is a nonprofit institution that helps improve policy and decisionmaking through research and analysis. RAND® is a registered trademark. RAND's publications do not necessarily reflect the opinions or policies of its research sponsors.

Published 2001 by RAND
1700 Main Street, P.O. Box 2138, Santa Monica, CA 90407-2138
1200 South Hayes Street, Arlington, VA 22202-5050
201 North Craig Street, Suite 102, Pittsburgh, PA 15213
RAND URL: http://www.rand.org/
To order RAND documents or to obtain additional information, contact Distribution Services: Telephone: (310) 451-7002; Fax: (310) 451-6915; Internet: order@rand.org

The Director of Officer and Enlisted Personnel Management asked the National Defense Research Institute (NDRI) to study enlisted force management to

- Assess how future enlisted requirements will continue to evolve, and

- Identify management changes that would provide a more effective and efficient match between future career enlisted inventories and requirements.

To understand how best to manage the enlisted force in the future, it is necessary to examine past and current management policies and practices and their effectiveness. We reviewed the literature on the enlisted force and distilled lessons learned from that literature regarding the effects of various policies aimed at recruiting, retaining, developing, promoting, and transitioning military personnel. This report presents an overview of the literature, with particular emphasis on its policy implications. The appendix presents a detailed, annotated bibliography of the literature. The work was completed in 1996 and is being published now for archival reasons. The findings and recommendations, which were based on the context at the time the research was completed, may no longer be relevant, although the review of the literature may be useful to researchers and policymakers interested in enlisted management.

This research was sponsored by the Under Secretary for Personnel and Readiness and was conducted within the Forces and Resources Policy Center of RAND's NDRI. NDRI is a federally funded research and development center sponsored by the Office of the Secretary of Defense, the Joint Staff, the Unified Commands, and the defense agencies.

CONTENTS

TABLES

PURPOSE OF PROJECT

The Director of Officer and Enlisted Personnel Management asked the National Defense Research Institute to undertake an enlisted force management study. The study has two primary objectives:

- To assess how future enlisted requirements will continue to evolve, and

- To identify management changes that would provide a more effective and efficient match between future career enlisted inventories and requirements.

The project comprises four major tasks.

The first task undertook a historical look at the enlisted force, cataloguing the changes in its shape, its size, and its composition (skill, paygrade, experience, gender, and race/ethnicity). It also provided a historical view of enlisted management policies and practices and other exogenous events that have shaped the enlisted force over the past century (Kirby and Thie, 1996).

The second task analyzed the experience of comparable organizations in managing their career and work forces; these organizations included private sector organizations, other militaries, and public sector organizations. The results of this task are documented in an annotated briefing (Levy, Thie, and Sollinger, unpublished).

Task 3 is to delineate a purpose and long-term objectives for future enlisted career management.

Task 4 is to outline alternative management practices that may be helpful to decisionmakers in achieving these long-term objectives.

To map task 4 into task 3, we need a systematic, comprehensive review of previous research on the enlisted force, with special emphasis on programs/policies/practices that have been tried or proposed and the effects of such practices with respect to the attainment of various objectives. Unfortunately, no such review appeared to exist; this report is an attempt to fill the gap. Although the original rationale for this task was to build a bridge between task 3 and task 4, we soon found that focusing on only those policies/practices that mapped into potential objectives was too limiting. Instead, we decided to survey the literature more broadly in an attempt to provide a review that would be useful beyond the confines of this particular project.

SCOPE OF THIS REPORT

We should be clear about what this report is and, perhaps more important, what it is not. This is not a critical literature review in the traditional sense. Nor have we attempted to be particularly selective; time and funding constraints did not allow us to analyze the studies in detail or combine them in a meta-analytic framework that might help determine their significance or importance. Instead, our focus is quite circumscribed. We focus on particular or promising policies/practices/alternatives that have had (important) effects on particular functions that an enlisted management system encompasses: accessing, developing and training, promoting, and transitioning. Because of this limited focus, we do not deal with the many other interesting findings of the different studies. However, the appendix does contain a comprehensive annotated bibliography of the research that has been conducted on the enlisted force over the past 15 years that the reader may find of interest. The report, however, cannot hope to do justice to the richness and complexity of the studies cited here. Instead, it should be viewed as a port of entry into the literature—a starting point that refers the reader to the source document for greater detail.

Despite the circumscribed nature of the discussion, we feel that the report—and the detailed appendix—is an important contribution to the literature and should prove useful as a reference work to other researchers and military manpower planners working on issues relating to the enlisted force.

FINDINGS

We had hoped originally to be able to categorize our survey of the literature according to the long-term objectives delineated for enlisted management. At the time of writing, however, we were still in the process of conducting focus groups with policymakers and researchers to elicit objectives and a long way from distilling them into a short but complete set of objectives for enlisted management. As a result, we decided to categorize major policy findings from our survey of the literature by four functional areas: accession, developing and training, promoting, and transitioning. This also had an added advantage: Categorizing the literature and policies/practices according to this typology appeared to highlight and illuminate the natural shifts in emphasis in enlisted force management over the last 20 years. Earlier, with the inception of the All-Volunteer Force (AVF), recruiting and retention appeared to receive the most attention; in the early 1980s and early 1990s, measuring and screening for performance came to the forefront; in the 1990s, transitioning people out of military service became important in the wake of the drawdown. Throughout, concern about compensation appears to run as a unifying thread. Compensation is clearly not a separate function in itself; it is an important (arguably the most important) tool in the panoply of human resource management policies that can be used to shape and manage the force to stated ends; it has also received the lion's share of attention in the literature.

In the bulleted list below, we group policies and findings into four broad areas: accession policies, training and developing policies, promotion policies, and transitioning policies. Within each area, findings are further grouped into subsets generally thematically or if an emphasis in the literature warrants it (for example, the Delayed Entry Program (DEP) has received a fair amount of attention and appears to hold some promise).

ACCESSION POLICIES

Delayed Entry Program (Table 3.1)

- DEP helps attract high-quality recruits and allows the services to smooth out fluctuations in recruitment and use of training facilities.

- Longer time in DEP increases DEP loss.

- Longer time in DEP reduces first-term attrition.

Recruiting Resources and Strategies (Table 3.2)

Recruiting Incentives:

- Two-year options—used by the Army—appear to attract high-quality recruits in general and into critical Military Occupational Specialties (MOSs) in particular (although earlier research found limited effects).

- Combined active/reserve options raise the flow of high-quality recruits to both the active and reserve forces.

- Bonuses offer a very flexible recruiting tool, with substantial effects on channeling recruits into hard-to-fill skill and term of enlistment choices.

- Educational benefits strongly influence high-quality enlistees.

- From a total force perspective (and judging by additional man-years generated), educational benefits are more cost-effective than bonuses, increases in number of recruiters, or advertising expenditures; this is primarily because those receiving such benefits tend to have higher rates of completion of the first term and an increased probability of joining the reserve. However, there are still questions regarding the loss of high-quality recruits to the active force and the effect of this on readiness and the problem of skill match on reserve entry. If prior service individuals are not matched in their active duty skill, then retraining costs will increase.

Recruiter Incentives:

- Additional recruiters increase the number of high-quality enlistments.

- General Accounting Office (GAO) (1994) points out that increases in the numbers of recruiters were not needed at that time because of smaller force sizes and inefficiencies in the current system and because the number of potential recruits is expected to increase at least until 2000; however, others feel that increases in the number of recruiters and advertising may be needed to counter reduced supply.

- Recruiter incentives need to be carefully structured because they can affect both the quantity and quality of enlistments; indeed, attempts are being made to link such incentives to the ability of recruits to graduate from training.

- Asch and Karoly (1993) also point to the importance of the job counselor in filling high-priority jobs; this was not given much attention before and may have led to an overestimate of the effect of the Army College Fund (ACF) on high-quality enlistments. Giving counselors five extra bonus points has about 2.5 times the effect on occupational fill rates as does offering applicants the ACF for entering specific occupations and is about 1.5 times as effective in increasing occupational fill rates as an enlistment bonus. However, Asch and Karoly point out that the ACF and enlistment bonus have positive market expansion effects (in addition to the skill-channeling effect) that need to be considered.

Other Recruiting-Related Strategies:

- Emphasizing training aspects of the military may help attract more enlistments; surveys suggest that this is a primary reason for interest in the military.

- Almost half of all enlistees come from the negative intention group; thus, it would be a mistake to focus only on those with positive intentions to enlist.

- Comparability between civilian and military pay is important in recruiting and retention and meeting personnel quality goals, although the civilian wage index for military pay adjustment needs to be carefully measured.

Screening for Enlistment, Attrition, and Job Performance
(Table 3.3)

- Raising Armed Forces Qualification Test (AFQT)[1] requirements for entry would screen out more blacks.

- A small set of factors can reveal a wide range of attrition risks, so screening on some of these might prove helpful in reducing attrition: educational attainment ($\downarrow$), expectations of higher education ($\downarrow$), unstable preemployment history ($\uparrow$), dependent status ($\uparrow$), women ($\uparrow$), minority status ($\downarrow$), AFQT (varies by MOS).

- Interpretation and enforcement of service policies have an important effect on attrition, so recruits of comparable quality have quite different attrition rates.

- Psychological screening may help identify those with a higher degree of organizational attachment, which has been found to have a significant effect on early attrition.

- Although the literature seems to show that recruiting high-quality enlistees is cost-effective, the Congressional Budget Office (CBO) (1986) raises questions of whether higher quality raises costs more than productivity after some point; for example, as the percentage of high school graduates rises from 76 to 90 percent, keeping constant the percentage of AFQT I–IIIA recruits,[2] the force's overall productivity increases by about 1.1 percent whereas manning costs rise by 1.8 percent (largely because of increases in enlistment bonuses); raising the percentage of AFQT I–IIIA recruits from 65 to 69 improves productivity by 0.25 percent while raising costs by 0.50 percent.

- The importance of the Armed Forces Vocational Aptitude Battery (ASVAB) lies in its ability to classify individuals into jobs for

---

[1]The AFQT measures the general trainability of recruits.

[2]Categories I–IIIA constitute the upper half of the AFQT distribution and are generally viewed as high quality.

which they are well-suited, to screen for ability to become skill qualified, and to predict later performance on the job; effective job classification results in higher productivity, higher reenlistment rates, and reduced training school attrition, and thus leads to lower costs.

- Larger pay differentials as grades increase raise performance, work effort, and retention of the most able individuals; however, this may have the downside of reducing cooperation and teamwork because skewed rewards may lead to a kind of cutthroat competition; in addition, intergrade differentials needed to elicit the same levels of effort and retention will be smaller, the larger the difference in rank-specific nonpecuniary rewards across grades.

Sources of Entry (Table 3.4)

- Some possibility exists for civilian substitution in technical areas but this would require careful implementation; having inexperienced higher-grade members may be particularly problematic in cases where the individuals must be deployable into combat, because successful deployment requires more than technical skills.

- Lateral entry from reserve components may be useful.

- Limiting lateral entry forces the military to "overstock"—that is, personnel must be hired not just for what they can do today but for what they can potentially do in the future; this requires setting entry pay higher than initial productivity might warrant to attract a large enough pool of entrants; at higher levels, productivity will exceed pay; however, senior personnel must be induced to leave to make room for younger, high-ability personnel and this requires constant thinning at some ranks.

TRAINING AND DEVELOPING POLICIES (Table 3.5)

- Some military occupations may be more amenable to civilian training (those with high training costs, high in civilian exchangeability) but there may be institutional barriers to implementation.

- Schoolhouse training appears to be more cost-effective than on-the-job (OJT) training in terms of productivity, but clearly there needs to be some optimal balance between the two; Wild and Orvis (1993) warn that significant shifts from Advanced Individual Training (AIT) to OJT (from the current status quo) could produce problems for readiness, particularly for complex and technical skills.

- Limited MOS consolidation could reduce training costs as could the use of distributed training, computer-based training, and use of simulations; in addition, these latter types of training could be beneficial in ongoing training: refresher courses or courses to teach new skills or impart new knowledge.

- Reports of successful implementation of gender-neutral performance standards are based largely on lack of complaints rather than on a careful assessment of whether recruits can meet the physical demands of their occupation.

- A range of innovative personnel practices increased job satisfaction and retention but show no real improvements in work quality, timeliness, or overall labor cost (Gilbert, 1991; Orvis, Hosek, and Mattock, 1993).

PROMOTION POLICIES (Table 3.6)

- Promotion tempo has large effects on reenlistment and previous studies incorrectly estimated the effects of pay, AFQT, and occupation on retention because they failed to account for promotion tempo.

- Reductions in the size of the noncommissioned officer (NCO) force would produce savings but effects on readiness are largely unknown.

- Up-or-out policies provide a means of increasing the exit rate of unqualified or mismatched workers; this is especially useful when ability and/or effort is observed imperfectly and unqualified workers cannot be induced to leave using voluntary means.

- It is costly to exceed career content limits both because a more senior force results in increased personnel costs in the form of military pay and retirement benefits and because not managing within targets creates cycles of peaks and valleys.

TRANSITIONING POLICIES

Retention Policies (Table 3.7)

- Family support programs may be very useful in increasing the level of spousal support for military life and increasing retention; spousal attitude is an important determinant of reenlist-ment/extension decisions, particularly among high-performing soldiers.

- The cost-effectiveness of Selective Reenlistment Bonuses (SRBs) has been shown by a number of studies; this is true across samples, across services, and across occupations; SRBs increase both the retention rate and the expected manyears of active duty service in a given occupation.

- Some studies have pointed to the need for more careful management of the SRB program; bonuses should be targeted to specialties with low reenlistment rates and high training costs.

- Reenlistment is also affected by a number of demographic and economic factors: high school graduates ($\downarrow$), minorities ($\uparrow$), perceived ease of finding alternative employment ($\downarrow$), unemployment ($\uparrow$), increase in military pay ($\uparrow$, but it varies by MOS and by quality groups), and promotion tempo ($\uparrow$).

Separating and Retirement Policies (Table 3.8)

- Voluntary separation payment plans need to be evaluated on the basis of cost, efficiency, and equitable treatment of those staying and those leaving; Grissmer, Eisenman, and Taylor (1995) suggest that hybrid plans that contain both a lump sum and an annuity would best meet these criteria.

- Retired pay provides a strong effort incentive and retention incentive to mid-career personnel; however, if increased retention of mid-career personnel reduces the promotion opportunities for junior personnel, it blunts the effort incentive of the latter.

- It is possible to revise the retirement system to increase productivity and maintain costs; this would require some increase in active duty pay and increased skewness of the pay table, along with separation benefits and some vesting at 10 years of service (Asch and Warner, 1994b).

- Increasing the portability of deferred benefits would be needed as a complement to increased use of lateral entries, both to make military service more attractive to potential lateral entrants (who might be looking to make further lateral moves later) and to provide fairness in advancement to service members (who need to be able to compete outside the military and for whom this would be very costly if deferred benefits were not portable); however, this alternative might have mixed benefits: attracting needed lateral entrants but decreasing productivity because of increased loss rates.

SOME FINAL THOUGHTS

It is important to keep in mind two caveats with respect to the policy implications derived from the literature review.

First, many of the studies reported here adopt a static approach: Changing x will produce outcome y but will fail to consider the interactions among parts of the system. Changing x might well change the expectations and behaviors of those both currently in and out of the military system and produce outcomes that far outweigh the original y. Most of the studies adopt a static approach and most focus on a specific policy or a particular issue. However, enlisted management needs to be viewed as a system, in which what is implemented in one area may have important effects on other areas. In addition, short-term effects need to be distinguished from long-term effects. Asch and Dertouzos (1994) provide a good example of this in their evaluation of the cost-effectiveness of educational benefits versus enlistment bonuses. They argue that comparisons based on recruiting effects alone provide only a partial picture of program effects and that what is needed is an examination of longer-term total force effects in terms of additional manpower generated (both active and reserve) rather than the immediate effects on enlistment, as is the more usual case. Such an approach leads to very different conclusions regarding the relative efficacy of one policy over another. When designing a system for the future, one needs to be aware of and think through the dynamic aspects of policies and changes in policies to avoid unwanted or undesired consequences.

Second, given the changing environment, past policies and practices may not work as well or as successfully as in the past to achieve organizational objectives.

The strategic human resource management approach adopted by the 8th Quadrennial Review of Military Compensation (QRMC) holds promise for the future and aligns itself with advances in personnel management in the civilian sector (DoD, 1998). In the most effective organizations, all systems—including the human resource management system—are aligned with strategy and various policies and practices are coordinated to work together as one system. The literature review has shown that many current service policies and practices are effective individually; what is important for the future, however, is to integrate these policies and practices to improve organizational performance.

ACKNOWLEDGMENTS

We are grateful to our sponsor, LTC Bradford Loo, the Assistant Director for Enlisted Policy, Office of Officer and Enlisted Personnel Management (OEPM) for his interest and support. The members of our project team (Harry Thie, Margaret Harrell, and Cliff Graf) provided helpful comments; Harry was particularly generous with his time, materials, and advice. We are grateful to our reviewer, James Hosek, for his constructive and thoughtful comments and to Susan Hosek for useful insights and her overall support and guidance. We thank Karshia Farrow for her patient and able assistance with the report and Patricia Bedrosian for her capable editing.

ACF	Army College Fund
ACT	American College Testing
ADAM	Aggregate Dynamic Analysis Model
AFEES	Armed Forces Entrance Examination Station
AFFS	Air Force Family Survey
AFQT	Armed Forces Qualification Test
AFSC	Air Force Specialty Code
AIT	Advanced Individual Training
ALC	Air Logistics Center
ALEC	Aggregate Lifecycle Effectiveness and Cost model
AMI	Apprentice Mechanic Initiative
APF	Appropriated Funds
ASVAB	Armed Services Vocational Aptitude Battery
ATC	Air Training Command
AVF	All-Volunteer Force
BMT	Basic Military Training
CBO	Congressional Budget Office
CBT	Computer-Based Training
CJR	Career Job Reservation

CL	Comparison Level
CLalt	Comparison Level for Alternatives
CNA	Center for Naval Analyses
DECI	Defense Employment Cost Index
DEP	Delayed Entry Program
DMDC	Defense Manpower Data Center
DoD	Department of Defense
ECI	Employment Cost Index
EFMS	Enlisted Force Management System
ETS	Expiration of Term of Service
EXPO	Experimental Civilian Personnel Office project
FAST	Fundamental Applied Skills Training
FBCT	Field-Based Cross Training
GAO	General Accounting Office
GED	General Equivalency Diploma
HRM	Human Resources Management
IRR	Individual Ready Reserve
ISC	Interservice Discharge Code
IST	Initial Skill Training
MEOA	Military Equal Opportunity Assessment
MEP	Military Entrance Processing Station
MORE	Multiple Option Recruiting Experiment
MOS	Military Occupational Specialty
MPT	Manpower, Personnel, and Training
NCO	Noncommissioned Officer
NEOSH	Navy Equal Opportunity/Sexual Harassment survey

NF	Nuclear Field
NLS	National Longitudinal Survey
NPS	Navy-Wide Personnel Survey
NLSY	National Longitudinal Survey—Youth
OEPM	Officer and Enlisted Personnel Management
OID	Organizational Identification
OJT	On-the-Job (training)
PCS	Permanent Change of Station
POF	Programmed Objective Force
QMM	Qualified Man-Month
QRMC	Quadrennial Review of Military Compensation
RMC	Regular Military Compensation
SQT	Skill Qualification Test
SR	Selected Reserve
SRB	Selective Reenlistment Bonus
SSB	Special Separation Bonus
TADSS	Training Aids, Devices, Simulators, Simulations
TEB	Targeted Enlistment Bonus
TIG	Time in Grade
TIS	Time in Service
TOE	Term of Enlistment
TPR	Trained Personnel Requirement
TQM	Total Quality Management
USAREUR	United States Army Europe
USMC	United Stated Marine Corps
VEAP	Veterans' Educational Assistance Program

VSI	Voluntary Separation Incentive
YATS	Youth Attitude Tracking Study
YOS	Years of Service

Chapter One

INTRODUCTION

ENLISTED FORCE MANAGEMENT[1]

Enlisted force management is concerned with meeting national military manpower requirements with the nation's citizens. It attempts to balance the demand for enlisted personnel as determined by the requirements process with the supply of enlisted personnel in a cost-effective manner. As such, it is concerned with questions such as:

- Size: How large a force in peacetime; how to increase it in wartime?

- Grade, skill, and experience composition: What skills are needed; how much experience is needed; how many are needed at various levels of supervision?

- Cost: What will it cost; are there tradeoffs; how best to procure and enter people into military service (accessing)?

- How best to train and experience (developing; assigning)?

- At what rate to advance in rank/grade when qualified (promoting)?

- How much to pay and in what form to make payment (compensating)?

---

[1]This section is taken from Kirby and Thie (1996).

1

- How best to send people from military service (separating, retir-ing, or transitioning)?

HISTORICAL BACKGROUND[2]

Although "careers" for officers have been the subject of debate in the United States for over 200 years, the idea of careers for enlisted members of the military services is a relatively recent development. Historically, few enlisted personnel made service a lifetime occupa-tion, as peacetime forces were small and pay was low. When large forces were needed, volunteers or conscripts were trained and used. After World War II, the Cold War resulted in large standing forces and periods of conscription; because pay remained low, fewer than 5 per-cent of entrants continued to retirement. However, since the begin-ning of the All Volunteer Force (AVF) in 1973, career considerations have come to the fore. The current active enlisted force comprises 1.35 million members, of whom nearly 800,000 are considered "careerists" in that they have five or more years of service. The aver-age time in service of the entire enlisted force is about eight years. Moreover, about 15 percent of new entrants are expected to continue to retirement.

The grade, skill, and experience composition of enlisted require-ments has also changed significantly over the years as well. For ex-ample, before World War II, light infantry supported by horse cavalry and horse-drawn artillery was the norm for the Army. World War II, Korea, and Vietnam saw significant increases in the use of mecha-nized forces and air power, although light infantry still dominated the land battlefield. Today, light infantry represents less than 10 per-cent of the force, and even these units have become highly special-ized. Similar evolution in grade, skill, and experience composition trends has taken place in the other services.

How enlisted members are accessed, trained, promoted, and transi-tioned has also changed significantly over the years. A high school educated entrant with high training aptitude is now the norm. Initial skill training may be as long as two years depending on the skill. Advanced skill courses and leadership education are standard.

---

[2]See Kirby and Thie (1996) for a detailed history of enlisted force management.

Promotions are more centralized and skill-based. Rules affecting continuation into the career force and to retirement have been imposed. Given the constrained fiscal environment and downsized forces, we need to ensure a balanced, ready, and cost-efficient enlisted force.

PURPOSE OF THE PROJECT AND THIS REPORT

Recognizing that the evolution of the enlisted force has substantial implications for managing it, the Director of Officer and Enlisted Personnel Management asked the National Defense Research Institute to undertake an enlisted force management study. The study has two primary objectives:

- To assess how future enlisted requirements will continue to evolve, and

- To identify management changes that would provide a more effective and efficient match between future career enlisted inventories and requirements.

The project comprises four major tasks.

The first task undertook a historical look at the enlisted force, cataloguing the changes in its shape, its size, and its composition (skill, paygrade, experience, gender, and race/ethnicity). It also provided a historical view of enlisted management policies and practices dealing with recruiting, developing, promoting, compensating, and separating enlisted personnel, and other exogenous events that have shaped the enlisted force over the past century (Kirby and Thie, 1996).

The second analyzed the experience of comparable organizations in managing their career and work forces; these organizations included private sector organizations, other militaries, and public sector organizations. The results of this task are documented in an annotated briefing (Levy, Thie, and Sollinger, unpublished).

Task 3 is to delineate a purpose and long-term objectives for future enlisted career management. To do that, we needed to delineate how future requirements would change. We adopt a simple—but reasonable—assumption that future requirements with respect to the experience, skill, and grade mix of the enlisted force will mirror the trends of the last five to ten years.

Task 4 is to outline alternative career management practices using a framework of personnel functions similar to the framework developed in the future officer management research (Thie et al., 2001).

To map task 4 into task 3, we need a systematic, comprehensive review of previous research on the enlisted force, with special emphasis on programs/policies/practices that have been tried or proposed and the effects of such practices with respect to the attainment of various objectives. Unfortunately, no such review appeared to exist; this report is an attempt to fill the gap. Although the original rationale for this task was to build a bridge between task 3 and task 4, we soon found that focusing on only those policies/practices that mapped into potential objectives was too limiting. Instead, we decided to survey the literature more broadly in an attempt to provide a review that would be useful beyond the confines of this particular project.

We should be clear about what this report is and, perhaps more important, what it is not. This is not a critical literature review in the traditional sense. Nor have we attempted to be particularly selective; time and funding constraints did not allow us to analyze the studies in detail or combine them in a meta-analytic framework that might help determine their significance or importance. Instead, our focus, dictated by the decision analysis methodology adopted for the study, is quite circumscribed: The main body of the report focuses on particular policies/practices/alternatives that have had (important) effects on particular functions that an enlisted management system encompasses: accessing, developing and training, promoting, and transitioning. Because of this limited focus, we do not deal with the many other interesting findings of the different studies. However, the appendix does contain a comprehensive annotated bibliography of the research that has been conducted on the enlisted force over the past 15 years that the reader may find of interest. This serves as a port of entry into the literature.

Despite the circumscribed nature of the discussion, we feel that the report—and the detailed appendix—is an important contribution to the literature and should prove useful as a reference work for other researchers and military manpower planners working on issues relating to the enlisted force.

ORGANIZATION OF THE REPORT

Chapter Two examines the characteristics of the current enlisted management system and helps set the context for the discussion of policies and practices in Chapter Three. The review in Chapter Three is organized by functional areas: accessing, developing, promoting, and transitioning. Clearly, there is some overlap in the studies cited under each area. Only studies that had direct policy implications are referenced in the main body of the report. A more comprehensive review is contained in the appendix, which presents abstracts of the studies organized alphabetically.

Chapter Two

CHARACTERISTICS OF THE CURRENT ENLISTED PERSONNEL MANAGEMENT SYSTEM[1]

Our review of history and analysis of personnel practices of the services suggest certain defining characteristics of the current enlisted system that describe its members and reflect the current management processes. We thought it might be useful for those unfamiliar with the enlisted system and enlisted management to briefly describe the system as it exists today. This will help provide a context for the discussion of the programs/practices reviewed in the next chapter. The characteristics of the enlisted system are organized around personnel functions: accessing, developing, promoting, and transitioning.

ACCESSING

Primarily entry at Year 0 for those meeting enlistment screens. Although each service uses prior active service individuals to meet particular skill needs, the vast majority of entrants are nonprior service. All those who enter must meet rigorous enlistment criteria.

Fixed contract periods. New entrants and those who elect to stay (reenlistees) are given "contracts" for service of a certain duration. Currently, new entrants generally sign up for three- or four-year enlistment terms; however, in the past some services have had primarily two-year enlistments whereas others have had high proportions of five- or six-year enlistees, particularly in skills with long training

---

[1]This section is taken from Kirby and Thie (1996).

times. Moreover, all entrants have a military service obligation of eight years, which implies either service in the selected reserve at the end of their initial active duty commitment or availability to be recalled to service if needed in emergencies.

Acculturation and initial skill training of all entrants. Initial entry training consists of two portions. The first—"basic" or "boot" training—focuses on general military skills and serves the purpose of acculturating individuals in the attitudes, values, and beliefs of each service as well as screening out those who are not likely to succeed. The second (e.g., Advanced Individual Training (AIT) in the Army) provides sufficient knowledge and ability to perform entry-level duties in a particular skill in a military unit. The first generally lasts eight weeks; the second from about eight weeks to over a year depending on the skill.

Emphasis on quality at entrance. The enlisted force contains the highest percentage of entering high school graduates ever—95 percent—and those who are in the upper half of the aptitude distribution—over 70 percent AFQT Category I–IIIA. This not only represents the services' efforts to increase quality, it also suggests military service (and a military career) is seen as a desirable occupation in itself or as a training ground for a different career or as a foundation before college.

DEVELOPING

Rank in person. Although most civilian organizations compensate (and promote) individuals based on the work they are currently doing and their level of responsibility, the U.S. military promotes the individual based on past performance and future potential. Thus, they continue to compensate an individual at that level, independent of the position actually occupied or the work actually being done.

Top five ranks and/or those with five or more years of service constitute the career force. Some women and men enter the military service intending to make a career. For most, however, the decision is made after a trial period—by both the individual and the service. The decision to make the service a career is frequently made at completion of the initial enlistment or after promotion to grade E-5, events that normally occur between the third and fifth year of service.

Family focused. The enlisted force today contains more married personnel than ever before (over 55 percent), with the resultant needs and demand on the management system.

Experienced and mature. The desire of the military has been traditionally for a "young and vigorous" force. However, the current force is the oldest in years, with over half being older than 24 years, and most experienced in seniority in history, with close to 50 percent being in the top five grades.

Retraining. Newer, younger people cost less in wages and benefits but lack firm-specific knowledge that the military prizes. As a result, the military emphasizes retraining, as old skills and jobs become outmoded. It also, of course, trains new entrants to meet new needs.

PROMOTING

Promotion based on combination of requirements and budget. In theory, promotion is based on service requirements for particular grades and specialties. Often, however, promotion is influenced by budgets that lead to manpower or grade ceilings.

Compensation for seniority. The compensation system is tilted toward attained levels of seniority and years of service.

Military work emphasizes team performance although rewards are based on individual performance. Much military work—especially that in operating units—is premised on the performance of team tasks that contribute to mission success. However, the rewards system—promotion and compensation—is heavily dependent on individual accomplishment.[2]

---

[2]Team-based performance is also fraught with its own set of problems, so it is not clear whether a wholly team-based performance assessment would be superior to the current system.

TRANSITIONING

Selective entry but high turnover. Although the services all use pre-enlistment testing, they still lose about 30 percent of each cohort before completion of three years service, with over 10 percent lost during initial training.

Retention controls. Each service uses tenure in grade and performance points to control continuation in the service, although only a small proportion are generally ineligible to continue.

Retirement at 20 years. Among those who choose an enlisted career, about 35 percent retire after 20 years of service.

MANAGEMENT PRINCIPLES

Mix of uniformity and flexibility in policy. Little uniformity across services and/or skills exists in certain policy areas (e.g., promotion, retention), which means that the services have great flexibility to manage their enlisted work force as they deem best. In the interests of equity across the military, the Congress and the Department of Defense have imposed greater uniformity in other areas (enlistment screening, basic pay, promotion criteria, retirement).

Chapter Three

ENLISTED MANAGEMENT POLICIES AND PRACTICES: EVIDENCE FROM THE LITERATURE

It might be helpful at the outset to reiterate what this chapter aims to accomplish, and, perhaps more important, what it does not.

First, as mentioned above, this chapter is not intended as a comprehensive literature review; that is beyond the scope and time constraints of this project. Some of the papers are seminal in the field and represent theoretical and statistical advances. We cannot do justice to them here. Instead, in our review, we attempt to distill lessons learned from the literature with respect to enlisted management policies and practices and the ability of these policies and practices to achieve management objectives.

Second, we have limited our focus to roughly the last 15 years; the enormous amount of literature dealing with the transition to the AVF seemed neither relevant nor useful in guiding policies and practices in the current environment and thus is not included here.

Third, we have excluded studies focusing primarily on officers because officers are managed and compensated in very different ways than enlisted personnel and personnel problems (and thus policy options for dealing with them) are quite different for the two groups.

Fourth, our primary focus is on *military* rather than civilian literature for several reasons: (a) The literature on civilian human resource management policies and practices is vast and enormously varied and it is difficult to do full justice to it in this report, given our time and budget constraints; (b) much of the literature appears not to be directly relevant to managing the enlisted force; (c) there are other summaries of the civilian literature; for example, the summary of the civilian and military literature on turnover and retention compiled by the Navy Personnel Research and Development Center (Wilcove, Burch, Conroy, and Bruce, 1991) and the review conducted by the 8th Quadrennial Review of Military Compensation that informed both their model and the organizational designs presented in their final report (DoD, 1996); (d) in addition, the second task of our project was to examine comparable civilian sector organizations to see how they managed their work force and whether their management practices could be usefully applied to managing the enlisted force; the results are documented in a briefing by Levy, Thie, and Sollinger.

Fifth, we found it constraining and not entirely easy or helpful to group the evidence from the literature according to their effects on the achievement of potential objectives. Indeed, by the time of this report, we did not have a final, exhaustive, nonredundant, operable set of objectives. Instead, we chose to cast the review more broadly so as to be useful to a larger audience. Because of these reasons, we selected a life-cycle approach to categorization. Ultimately, enlisted force management is concerned with questions as:

- How best to procure and enter people into military service (accessing).

- How best to train them and give them experience (developing, assigning).

- At what rate to advance them in rank/grade when qualified (promoting).

- How best to remove people from military service (separating, retiring, or transitioning) (Kirby and Thie, 1996, pp. 3–4).[1]

Categorizing the literature and policies/practices according to this typology appears to highlight and illuminate the natural shifts in emphasis in enlisted force management over the last 20 years: Earlier, with the inception of the AVF, recruiting and retention appeared to receive the most attention; in the early 1980s and 1990s, measuring and screening for performance came to the forefront; in

---

[1]There is little written on requirements; as a result, we chose to exclude this topic from our report.

the 1990s, transitioning people became important in the wake of the drawdown. Throughout this period, concern about compensation appears to run as a unifying thread: how much, how much in cash versus in-kind benefits, the effectiveness of bonuses versus increases in base pay, the cost-effectiveness of bonuses versus other incentives such as educational benefits, and more recently, concern about vesting points and retirement benefits. Compensation is clearly not a separate function in itself; it is an important (arguably the most important) tool in the panoply of human resource management policies that can be used to shape and manage the force to stated ends; it has also received the lion's share of attention in the literature.

Sixth, we have chosen a tabular rather than a discursive format for presenting the review of policies and practices because it provides the information in a more succinct fashion; each table is organized thematically and lists policies/practices derived from the study, authors of the study and date published, information about the sample where available, and more detailed notes that may of interest to readers. In addition, the appendix provides a complete bibliography of the literature on the enlisted force. The bibliography contains short summaries of each study and keywords that allow cross-referencing and is organized alphabetically. Studies that did not appear to have a direct policy implication are not referenced in the body of the report but are included in the appendix.

ACCESSION POLICIES

Accession policies are described in Tables 3.1–3.4; these focus on the Delayed Entry Program (DEP); recruiting resources and strategies; screening for enlistment, attrition, and performance; and sources of entry respectively.

Delayed Entry Program (Table 3.1)

The DEP allows individuals who have signed enlistment contracts to delay reporting for active duty up to a period of 12 months. This allows services to smooth out the seasonal fluctuations in recruiting, attract high-quality recruits (particularly high school seniors), and maintain a steady use of training facilities. It may also increase recruiting productivity as DEP recruits provide referrals. However, some tradeoffs need to be balanced against the advantages offered by DEP. As Table 3.1 shows, length of time in the DEP is associated with higher DEP loss rates; such losses lead to empty training seats and perhaps wasted training resources if replacements are not immediately available. In addition, DEP loss requires expenditure of additional resources to fill those empty seats. Maintaining DEP pools requires recruiter time and effort to keep track of enlistees, provide them with activities to keep up their interest in military service, and help solve potential problems. However, studies of first-term attrition show that length of time in DEP lowers first-term attrition significantly, perhaps because it helps screen out those least sure of their decisions and so most likely to leave early in the term (Antel, Hosek, and Peterson, 1987). Manganaris and Phillips (1985) explicitly model the tradeoff between increased DEP loss and reduced attrition and suggest that length of DEP should vary by MOS:

- DEP helps attract high-quality recruits and allows the services to smooth out fluctuations in recruitment and use of training facilities.

- Longer time in DEP increases DEP loss.

- Longer time in DEP decreases first-term attrition.

Recruiting Resources and Strategies (Table 3.2)

We broadly divide recruiting resources and strategies into recruiting incentives, recruiter incentives, and other strategies.

Recruiting Incentives

- Two-year options—used by the Army—appear to attract high-quality recruits in general and into critical MOSs in particular (although earlier research found limited effects).

- Combined active/reserve options raise the flow of high-quality recruits to both the active and reserve forces.

- Bonuses offer a very flexible recruiting tool, with substantial effects on channeling recruits into hard-to-fill skills and term of enlistment choices.

- Educational benefits strongly influence high-quality enlistees.

- From a total force perspective (and judging by additional man-years generated), educational benefits are more cost-effective than bonuses, increases in number of recruiters, or advertising expenditures; this is primarily because of the higher rate of completion of the first term and the increased probability of joining the reserve. However, there are still questions regarding the loss of high-quality recruits to the active force and the effect of this on readiness and the question of skill match on reserve entry. If prior service individuals are not matched in their active duty skill, then retraining costs will increase.

Recruiter Incentives

- Additional recruiters increase the number of high-quality enlistments.

- GAO (1994) points out that increases in the number of recruiters are not needed because of smaller force sizes and inefficiencies in the current system and because the number in the potential recruit market is expected to increase at least until 2000; however, others feel that increases in the number of recruiters and in advertising may be needed to counter reduced supply.

- Recruiter incentives need to be carefully structured because they can affect both the quantity and quality of enlistments; indeed, attempts are being made to link such incentives to the ability of recruits to graduate from training.

- Asch and Karoly (1993) also point to the importance of the job counselor in filling high-priority jobs; this has not been given much attention before and may have led to an overestimate of the effect of the ACF on high-quality enlistments. Giving counselors five extra bonus points has about 2.5 times the effect on occupational fill rates as does offering applicants the ACF to enter that occupation and is about 1.5 times as effective in increasing occupational fill rates as an enlistment bonus. However, Asch and Karoly point out that the ACF and enlistment bonus have positive market expansion effects (in addition to the skill-channeling effect) that need to be considered.

Other Recruiting-Related Strategies

- Emphasizing training aspects of the military may help attract more enlistments; surveys suggest that this is a primary reason for interest in the military.

- Almost half of all enlistees come from the negative intention group; thus, it would be a mistake to focus only on those with positive intentions to enlist.

- Comparability between civilian and military pay is important in recruiting and retention and meeting personnel quality goals although the civilian wage index for military pay adjustment needs to be carefully measured.

Screening for Enlistment, Attrition, and Job Performance (Table 3.3)

A major focus of the literature over the past 15 years or more has been developing criteria that would allow the military to select those most likely to join, to stay, to complete the first term, and to perform well. These criteria have ranged from demographic and economic characteristics (race/ethnicity, gender, age, marital status, labor force participation), intention to enlist, educational attainment, and performance on the Armed Forces Qualifying Test. Many studies have pointed to the relationship between enlistment, attrition, and the state of the economy; Buddin (1988) also highlighted the effect that service policies have on attrition, which can often negate the effect of raising quality among entry cohorts. Throughout this period, economists pointed to the importance of military pay in meeting enlistment and end-strength goals, and intragrade pay skewness in improving effort and performance. In addition, the importance of AFQT scores as a screening mechanism is underscored both for completion of the first term and for performance on the job (particularly in the more recent literature). Some of the major findings are listed below:

- Raising AFQT requirements for entry would screen out more blacks.

- A small set of factors can reveal wide range of attrition risks, so screening on some of these might prove helpful in reducing attrition: educational attainment ($\downarrow$), expectations of higher education ($\downarrow$), unstable preemployment history ($\uparrow$), dependent status ($\uparrow$), women ($\uparrow$), minority status ($\downarrow$), AFQT (varies by MOS).

- Interpretation and enforcement of service policies have an important effect on attrition, so recruits of comparable quality have quite different attrition rates.

- Psychological screening may help identify those with a higher degree of organizational attachment,[2] which has been found to have a significant effect on early attrition.

- Although the literature seems to show that recruiting high-quality enlistees is cost-effective, CBO (1986) raises questions of whether higher quality raises costs more than productivity after some point; for example, as the percentage of high school graduates rises from 76 to 90 percent, keeping constant the percentage of AFQT I–IIIA recruits, the force's overall productivity increases by about 1.1 percent whereas manning costs rise by 1.8 percent (largely because of increases in enlistment bonuses); raising the percentage of AFQT I–IIIA recruits from 65 to 69 improves productivity by 0.25 percent while raising costs by 0.50 percent.

- The importance of ASVAB lies in its ability to classify individuals into jobs for which they are well-suited, to screen for ability to become skill-qualified, and to predict later performance on the job; effective job classification results in higher productivity, higher reenlistment rates, and reduced training school attrition, thus leading to lower costs.

- Larger pay differentials as grades increase raise performance, work effort, and retention of the most able individuals; however, this may have the downside of reducing cooperation and team-

---

[2]Organizational attachment is defined as "a specific form of social identification in which people define themselves in terms of their membership in a particular organization; . . . in a real sense, people who identify may see themselves as *personifying* the organization." (Mael and Ashforth, 1995, pp. 311–312).

work because skewed rewards may lead to a kind of cutthroat competition;[3] in addition, intergrade differentials needed to elicit the same levels of effort and retention will be smaller the larger the difference in rank-specific nonpecuniary rewards across grades.

Sources of Entry (Table 3.4)

The means of accessing people has been primarily entry at Year 0 for those meeting enlistment screens, although each service uses prior service individuals to meet particular skill needs. As Robbert et al. (1997) point out, "An externalized labor market has the potential to decrease HRM costs. Moreover, as existing non-military technologies are imported into military usage at an ever increasing rate, the services may find it necessary to externalize their military labor markets through direct appointment at middle or upper grades of individuals with needed technical skills." (pp. 85–86). However, privatization, outsourcing, or civilianization of these jobs are other alternatives to lateral entry that are increasingly being considered.

Very little attention has been devoted in the literature to the issue of lateral entry and what changes might be necessary for this to work efficiently. Asch and Warner (1994a) outline the costs and consequences to the organization of limiting lateral entry, although it must be emphasized that little is known about lateral entry supply curves. The possibility of some substitution of civilians for military personnel may be possible in some technical areas, but as Robbert et al. (1997) point out, there are concerns that need to be addressed in order for this to be successful: perceptions of unfairness among the eligible military members, portability of deferred benefits, and difficulties on the part of civilians in adjusting to the military culture. A more promising avenue is transfer of reservists to the active force, but there is little direct evidence regarding the cost-effectiveness of such a policy.

- Some possibility exists for civilian substitution in technical areas, but this would require careful implementation; having inexperi-

---

[3]This has not been tested empirically.

enced higher-grade members may be particularly problematic in cases where the individuals must be deployable into combat, because successful deployment requires more than technical skills.

- Lateral entry from reserve components may be useful.

- Limiting lateral entry forces the military to "overstock"—that is, personnel must be hired not just for what they can do today but for what they can potentially do in the future; this requires setting entry pay higher than initial productivity might warrant to attract a large enough pool of entrants; at higher levels, productivity will exceed pay; however, senior personnel must be induced to leave to make room for younger, high-ability personnel and this requires constant thinning at some ranks. Assumptions regarding the lateral entry supply curve and the role of the reserves in this process need to be examined.

TRAINING AND DEVELOPING POLICIES (Table 3.5)

Studies focusing on training address such questions as whether civilian training is an appropriate alternative to military training, whether schoolhouse training can be replaced by on-the-job (OJT) training, and how overall costs can be reduced. There is a dearth of literature examining development of the enlisted force, although this is indirectly addressed in studies examining promotion and promotion tempo. This may be partially explained by the fact that enlisted jobs are not seen as "careers," in the manner of officers, and thus lack a well-defined career path. Recently, however, this is receiving greater attention as the services attempt to look at NCO leader development. The major findings in the area of training and developing are summarized below:

- Some military occupations may be more amenable to civilian training (those with high training costs, high in civilian exchangeability), but there may be institutional barriers to implementation.

- Schoolhouse training appears to be more cost-effective than OJT in terms of productivity, but clearly there needs to be some optimal balance between the two; Wild and Orvis (1993) warn that significant shifts from AIT to OJT (from the current status quo) could produce problems for readiness, particularly for complex and technical skills.

- Limited MOS consolidation could reduce training costs, as could the use of distributed training, computer-based training, and use of simulations; in addition, these latter types of training could be beneficial in ongoing training: refresher courses or courses to teach new skills or impart new knowledge.

- Reports of successful implementation of gender-neutral performance standards are based largely on lack of complaints rather than on a careful assessment of whether recruits can meet the physical demands of their occupation.

- A range of innovative personnel practices increased job satisfaction and retention but show no real improvements in work quality, timeliness, or overall labor cost (Gilbert, 1991; Orvis, Hosek, and Mattock, 1993).

PROMOTION POLICIES (Table 3.6)

As we pointed out in Chapter Two, promotion is based on a combination of requirements and budget. With most of the service, promotion rates differ across specialties (although sometimes not in the expected direction—see Kirby and Thie (1996), pp. 25–60, for a history of DoD and the services' largely unsuccessful attempts to control grade and seniority over the last 40 years).

As with training and development, promotion has received short shrift in the literature. A notable exception is the work done by Buddin et al. (1992), which shows that promotion tempo has large effects on reenlistment and that previous studies incorrectly estimated the effects of pay, AFQT, and occupation on retention because they omitted to account for promotion tempo. Current plans call for reductions in the proportion of NCOs in the Army; while this would produce savings, the effects on readiness are largely unknown. Asch and Warner (1994a) highlight the importance of up-or-out policies as providing a means of increasing the exit rate of unqualified or mismatched workers. GAO (1991) underscores the costs of exceeding career content limits.

- Promotion tempo has large effects on reenlistment and previous studies incorrectly estimated the effects of pay, AFQT, and occupation on retention because they omitted to account for promotion tempo.

- Reductions in the size of the NCO force would produce savings but effects on readiness are largely unknown.

- Up-or-out policies provide a means of increasing the exit rate of unqualified or mismatched workers; this is specially useful when ability and/or effort is observed imperfectly and unqualified workers cannot be induced to leave using voluntary means.

- It is costly to exceed career content limits both because a more senior force results in increased personnel costs in the form of military pay and retirement benefits and because not managing within targets creates cycles of peaks and valleys.

TRANSITIONING POLICIES

Separating people from military service occurs at all points in the military career but from a policy perspective, it is reenlistment at the end of the first and second terms and retirement that are of interest. Along with recruiting, retention (particularly reenlistment at the first term) has received the lion's share of attention in the last 15 years. Because of this, we have divided transitioning policies into two broad categories: retention policies, dealing mostly with reenlistment or extension, and separation/retirement policies dealing with issues of downsizing and the retirement system. The former are described in Table 3.7; the latter in Table 3.8.

RETENTION POLICIES (Table 3.7)

Most of the attention has focused on the efficacy of the Selective Reenlistment Bonus (SRB) in retaining high-quality individuals, in filling hard-to-fill specialties, and choice of term of reenlistment, along with the effect of pay and unemployment. Some studies have underscored the importance of spousal attitudes and job satisfaction in the reenlistment/extension decision, whereas others have examined how the reenlistment rate varies across demographic group. We list the major findings below:

- Family support programs may be very useful in increasing the level of spousal support for military life and increasing retention; spousal attitude is an important determinant of reenlist-ment/extension decisions, particularly among high-performing soldiers (Lewis, 1985; Lakhani et al., 1985; Rakoff, Griffith, and Zarkin, 1994).

- The cost-effectiveness of SRBs has been shown by a number of studies; this is true across samples, across services, and across occupations; SRBs increase both the retention rate and the ex-pected manyears of active duty service in a given occupation.

- Some studies have pointed to the need for more careful man-agement of the SRB program.

- SRBs should be varied by MOS.

- Reenlistment is also affected by a number of demographic and economic factors: high school graduates ($\downarrow$), minorities ($\uparrow$), perceived ease of finding alternative employment ($\downarrow$), unemployment ($\uparrow$), increase in military pay ($\uparrow$), although this varies by MOS and by quality groups, and promotion tempo ($\uparrow$).

SEPARATING/RETIREMENT POLICIES

The early 1990s saw a period of great concern regarding the design and implementation of voluntary separation payment plans for en-listed personnel, but retirement policies have been hotly debated in the services and in Congress ever since the start of AVF. Compensation is the subject of quadrennial reviews and Congress has sought to reform the military retirement system at different points. Asch and Warner (1994a) discuss the role of retired pay:

- As deferred compensation, retired pay can encourage both effort and retention of younger personnel.

- It provides individuals with an incentive to separate voluntarily.

- As a replacement for mandatory retirement, it is both ex ante and ex post incentive-compatible (p. 118).

Some of the recommendations in the literature regarding separa-tion/retirement policies include:

- Voluntary separation payment plans need to be evaluated on the basis of cost, efficiency, and equitable treatment of those staying and those leaving; Grissmer, Eisenman, and Taylor (1995) suggest that hybrid plans that contain both a lump sum and an annuity would best meet these criteria.

- Retired pay provides a strong effort incentive and retention incentive to mid-career personnel; however, if increased retention of mid-career personnel reduces the promotion opportunities for junior personnel, it blunts the effort incentive of the latter.

- It is possible to revise the retirement system to increase productivity and maintain costs; this would require some increase in active duty pay and increased skewness of the pay table, along with separation benefits and some vesting at 10 years of service (Asch and Warner, 1994b).

- Increasing the portability of deferred benefits would be needed as a complement to increased use of lateral entries, both to make military service more attractive to potential lateral entrants (who might be looking to make further lateral moves later) and to provide fairness in advancement to service members (who need to be able to compete outside the military and for whom this would be very costly if deferred benefits were not portable); however, this alternative might have mixed benefits: attracting needed lateral entrants but decreasing productivity because of increased loss rates.

The remainder of this chapter presents tables describing the studies from which these policies/programs/practices were drawn.[4]

---

[4]We remind readers that studies that examine several issues may be listed under several different headings. This was done to make each subsection as complete as possible.

Table 3.1

Delayed Entry Program (DEP)

Effect of Policy/Practice	Reference	Data	Notes
Effective DEP management may help reduce DEP loss	Zimmerman, Zimmerman, and Lathrop (1985)	Survey of Army enlistees participating in DEP during FY84 (n=1,000)	Major findings were that individuals who separated while in DEP tended to be dissatisfied with their occupational assignment, wanted to attend school, found a civilian job, experienced a change in attitude; positive experiences during DEP, development of group cohesiveness, and greater weight to applicant preferences in job assignment are important in retaining DEP participants
Longer time in DEP increases DEP loss	Celeste (1985)	Signed contracts written between October 1980 and March 1983 (n=375,341)	Loss rates were higher for AFQT levels II and IIIA compared to those of AFQT levels I and IIIB, for GEDs and nonhigh school graduates compared to seniors and high school graduates, for women compared to men; findings were somewhat inconsistent across the fiscal years
	Kearl and Nelson (1992)	Individuals who signed contracts to enlist in the Army during FY86–FY87 and were in DEP (n=241,420)	A number of economic and demographic characteristics influence DEP loss: high unemployment and high military/civilian earnings, ACF, and enlistment bonuses reduce DEP loss; high school graduates (compared to seniors), blacks, males, younger recruits (particularly among seniors), and those with higher AFQT scores have lower loss probabilities; screening for age/presence of dependents, providing enlistment benefits, and academic counseling may help reduce DEP loss

Table 3.1 (continued)

Effect of Policy/Practice	Reference	Data	Notes
Longer time in DEP reduces first-term attrition	Buddin (1981)	FY75 nonprior enlisted accession cohort	Buddin examined post-training attrition as a function of individual background characteristics and recruit service experiences (including post-training duty location, MOS, career turbulence, and participation in DEP); recruits who spent more than three months in DEP had 5 to 10 percent lower attrition probability than nonparticipants
	Manganaris and Schmitz (1985)	FY81 enlisted accession cohort; 13 MOSs with n>500 selected (n=40,776)	The sensitivity of attrition to DEP participation varies by MOS but substantial savings could be achieved by lengthening average DEP by 5 percent; however, maintaining individuals in DEP is costly
	Antel, Hosek, and Peterson (1987)	1979 DoD Survey of Personnel Entering the Military Service pooled with nonenlistee data from the National Longitudinal Study of Labor Market Behavior Youth Survey; recruits tracked through 1984	Effect of months in DEP significant in reducing 35-month attrition of seniors and graduates and six-month attrition of seniors (but not graduates)
Tradeoff between minimizing DEP loss and minimizing attrition	Manganaris and Phillips (1985)	FY83 contracts; 12 MOSs (n=66,000)	DEP policy should be MOS-specific; increasing DEP lengths is cost-effective

Table 3.2

Recruiting Resources and Strategies

Effect of Policy/Practice	Reference	Data	Notes
		Recruiting Incentives	
Educational benefits dominate bonuses as a recruiting option and compare favorably with other policy options, such as increases in recruiters or advertising expenditures	Asch and Dertouzos (1994)	Recruits who joined the Army during the Enlistment Bonus Test and the Educational Assistance Test Program in the 1980s tracked forward through FY88; various cost factors	Cost-effectiveness comparisons based on recruiting effects alone provide only a partial picture of program effects; what is needed is an examination of total force effects including attrition, retention, and reserve component accession rates of personnel and of the active manyears contributed; the relative cost-effectiveness should be measured in terms of additional manpower generated; educational benefits lead to (a) a significant increase in enlistments and the rate of completion of the first term, (b) an increase in the likelihood of separating at the end of the first-term, (c) an increase in the probability of reserve entry, (d) a reduction in actual cost of the program relative to bonuses because payments are deferred several years, and some eligible individuals choose not to utilize the full amount available; on a cost-per-manyear basis, educational benefits dominate bonuses; increasing recruiters or advertising expenditures costs the same per additional enlistment as educational benefits but does not generate increased flow to the reserve components; thus, from a total force perspective, educational benefits emerge as the preferred policy alternative

Table 3.2 (continued)

Effect of Policy/Practice	Reference	Data	Notes
Combined active/reserve enlistment offers can raise the number of high-quality recruits in both the active and reserve forces	Buddin (1991); Buddin and Polich (1990)	Participants in the 2+2+4 experiment compared to control groups, 1989–1990 (n ≈ 6,000 participants in experiment)	The Army College Fund (ACF) was expanded to cover certain types of enlistments in selected noncombat occupational specialties under a new option, called "2+2+4," which consisted of a two-year active duty tour, an additional two-year commitment in the Selected Reserve and approximately four years in the Individual Ready Reserve (IRR); the program expanded the market for high-quality recruits by about 3 percent, did not shift a large number of recruits away from longer terms of service, and channeled recruits into hard-to-fill noncombat skills
Bonuses are very flexible and have substantial effects on recruiting, enlistees' skill, and term of enlistment choices	Polich, Dertouzos, and Press (1986)	Participants in the Army's Enlistment Bonus Experiment from 1982–1984 and base period data	Two new bonuses were offered to high-quality enlistees enlisting in selected combat specialties: an increased bonus for four-year enlistment (from $5,000 to $8,000) and a new $4,000 bonus for a three-year enlistment; produced a statistically significant market expansion effect and large effects on skill and term of service choices; the expanded bonuses attracted more recruits and lengthened their average term of commitment

Table 3.2 (continued)

Effect of Policy/Practice	Reference	Data	Notes
Targeted Enlistment Bonuses (TEB) can help smooth seasonal patterns in accessions for some specialties and save money relative to a nontargeted enlistment bonus	Cooke (1987)	Nuclear field recruits in 1983–1987 (seasonal and monthly data)	The TEB for Nuclear Field (NF) recruits in the Navy was an 18-month experiment that offered lower bonuses for summer than for spring accessions and was successful in achieving the desired change in seasonal pattern of accessions; the program offered savings in bonus expenditures relative to a nontargeted enlistment bonus; the size of the NF DEP grew whereas the overall DEP remained constant and average length of time in DEP increased; quality remained constant
Educational benefits strongly influence enlistment decisions of high-quality soldiers as does the two-year enlistment option	Gray (1987)	17–24 year old nonprior service high school graduate enlistees responding to the 1985 New Recruit Survey (n=5,752)	Those scoring in the upper half of the test score distribution were more likely to report that they would not have signed up for the same job if it did not qualify for the ACF or another cash bonus, and would not have enlisted if a two-year option was not available; one drawback is that such options and bonuses encourage soldiers to leave at the end of their enlistment term; might want to consider some options where options to stay in the Army or leave to gain further education are not mutually exclusive

Table 3.2 (continued)

Effect of Policy/	Reference	Data	Notes
Enlistment incentives—two-year term of enlistment, expanded postservice educational benefits, and an IRR options—have limited effects on increasing high-quality enlistments to hard-to-fill specialties	Haggstrom, Blaschke, Chow, and Lisowski (1981)	Monthly counts of enlistment contracts for each Armed Forces Entrance Examination Station (AFEES), 1978–1979	In early 1979, the services began the Multiple Option Recruiting Experiment (MORE) to test the attractiveness of new enlistment incentives; in addition to two-year enlistment option and increased educational benefits, the Army tested an option permitting recruits to choose reserve duty in lieu of active service after initial training (the IRR option); for the most part, MORE options were restricted to high-quality recruits in selected (hard-to-fill) specialties; none of the options yielded sizable enlistment responses; the "VEAP kickers" (additional money for education added to the current Veterans' Educational Assistance Program or VEAP) showed a modest effect but might have a detrimental effect on retention at expiration of term of service (ETS); the IRR option showed some promise for recruiting lower-quality males into combat arms
Those with higher educational expectations are more likely to enlist in the military so educational benefits might prove a good recruiting tool; pay was not a strong factor in enlistment decisions	Borus and Kim (1985)	National Longitudinal Survey of Youth, 1979–1981 (n ≈ 12,600)	Among high school graduates who were not attending college, those with higher educational aspirations preferred military to civilian alternatives; pay was neither a strong incentive or disincentive to enlistment (1979 was a poor year for recruiting); recruits entering the military were comparable in quality (or somewhat higher) than those employed full-time; intention to enlist is much lower in whites than minorities; among whites, desire for occupational training was a major motivation for entering the military

Table 3.2 (continued)

Effect of Policy/Practice	Reference	Data	Notes
The most cost-effective policies for force shaping and controlling flows into specific specialties are enlistment rate adjustment, Zone A bonuses, and Career Job Reservations; less effective are Zone C bonuses, early releases, and unnecessary retraining for changes of specialties	Rydell (1987)	Uses Air Force enlisted cohorts to develop a model for analyzing the cost and effectiveness of management policy	The primary purpose of this research is the development of the Aggregate Lifecycle Effectiveness and Cost (ALEC) model for ongoing use to evaluate alternative force management actions; the analysis of particular policies is offered as illustration with results that are more suggestive than definitive
Two-year enlistment option and the ACF appear to increase enlistment of top prospects into critical MOSs	Elig, Johnson, Gade, and Hertzbach (1984)	Surveys of 25,000 Army recruits in 1982 and 1983	Every 100 two-year enlistments give a net gain of 71 years of service; every 100 ACF enlistments are estimated to bring in 35 high-quality males who otherwise would not have enlisted and to give 13 MOS shifts by other high-quality males

Table 3.2 (continued)

Effect of Policy/Practice	Reference	Data	Notes
Enlistment bonuses have increased accessions into nuclear-related specialties and completion of skill training	Quester and Jeffries (1985)	Navy, enlisted, NF obligors from 1974 through 1985	Increasing the enlistment bonus by $1,000 adds 30 new obligors to the DEP, increases total shipments by about 18 recruits, and reduces the Navy's NF recruit shortage by about 18
The Army College Fund is roughly equal to enlistment bonuses in cost-effectiveness	Hogan, Smith, and Sylvester (1991)	10 percent sample of the Army FY83 recruit cohort	Army College Fund has no effect on attrition, decreases retention by 10 percent after two-year contracts and by 5 to 6 percent after three- to four-year contracts
Recruiter Incentives			
Recruiter incentives can affect both the quantity and quality of enlistments	Dertouzos (1984; 1985)	Monthly Army enlistment data for 1980–1981	Ignoring recruiter behavior and choices can lead to incorrect estimates of the effects of economic changes and recruiting resource expenditures; recruiters are motivated to attain quotas for both high- and low-quality enlistments but have few incentives to exceed them; thus, once quotas are met, additional resources or a rise in the supply of enlistments may not translate into increased number of contracts
Additional recruiters increase the number of high-quality enlistments	Brown (1984)	Number of monthly contracts signed by nonprior service male Army recruits, FY75–FY82	The findings indicate that increases in other services' recruiters, holding the number of Army recruiters constant, reduced the flow of applicants to the Army; the effect of national media advertising was positive but unstable across specifications; local media advertising had little effect

Table 3.2 (continued)

Effect of Policy/Practice	Reference	Data	Notes
	Hosek, Fernandez, and Grissmer (1984)	Historical 1977–1982 data on enlistments, retention, and variables affecting them used to project active enlisted supply, 1983–1990	Earlier studies had shown that adding 1 percent to the size of the recruiting force would increase high-quality male enlistments by 0.5–0.8 percent; however, the Army's recruiting success in the early 1980s was achieved largely through restructuring the recruiter incentive system via the introduction of the "mission box," which defined missions broken out by various categories (such as graduate/nongraduate, and AFQT category) rather than a total quota, as was previously the case
Improve recruiter incentives to encourage better screening of applicants	GAO (1997)		The study reports that the Navy Recruiting Command subtracts a percentage of incentive points from its recruiters when enlistees fail to graduate from training and adds a smaller number of incentive points for those who do; modifications to this policy being considered are to deduct total incentive points for all recruits, regardless of reason, who separate within the first 30 days of recruit training; the Marine Corps has, for several years, provided its recruiting units the flexibility to tie recruiters' incentive system to enlistees' successful completion of basic training; the Air Force and Army have expressed concern that recruiters should not be held accountable for what happens at the training base

Table 3.2 (continued)

Effect of Policy/Practice	Reference	Data	Notes
Increases in numbers of recruiters are not needed because of smaller future military force size and inefficiencies in the current recruiting system	GAO (1994)	Review of military recruiting system and recent literature	GAO feels that the services have overstated the potential recruiting challenges: The number of potential recruits is expected to increase at least until 2000 whereas the number of recruits needed has decreased; in addition, the declining propensity of youth to enlist may not be a good indicator of behavior, since half of the enlistees come from groups with negative intentions to join the military; the services have found ways to reduce first-term attrition but these have not been adopted DoD-wide or even throughout any of the services; half of the current 6,000 recruiting offices supply only 13.5 percent of enlistees; restructuring current organizational structures is needed to eliminate layers of management and cut costs
Advertising, educational benefits, and the number of recruiters are cost-effective and flexible resources when increased to counter decreasing supply and difficulties in converting potential supply into recruits	Orvis, Sastry, and McDonald (1996)	Youth Attitude Tracking Survey, DoD Recruiter Survey, High School ASVAB database, and the "Monitoring the Future" survey of high school seniors	Changes in use of recruiting resources should be tailored to meet specific shortages, either in specific jobs or across many specialties; recent patterns show declining propensity for youth to enlist and continuing difficulties in converting potential supply into enlistments

Table 3.2 (continued)

Effect of Policy/Practice	Reference	Data	Notes
Freeman Plan of recruiter incentives does have an effect on recruiter performance, though possibly not in ways that best benefit the Navy	Asch (1990)	Data from the Chicago Navy Recruiting District for five months in FY86	Adjustments to the Freeman Plan are suggested; raising the point differential for high-quality recruits would better focus recruiters on these prospects. Shortening award cycles would help smooth levels of effort within production cycles
Job counselors have an important role in filling high-priority jobs and occupations; counselor incentive plans need to be carefully structured	Asch and Karoly (1993)	Enlistments into high-priority jobs and occupation enlistments from FY86–FY89; data on counselor incentives plans	Failing to account for the counselors' role leads to overestimates of the effect of the ACF on high-quality enlistments; offering counselor incentives has dramatic effects on the ability of the Army to fill occupations and priority jobs (offering counselors five more incentive plan points is 2.5 times more effective than educational benefits in filling an occupation and 1.5 times more effective than offering an enlistment bonus) and is much more cost-effective in skill-channeling than bonuses or educational benefits; rewarding counselors on the basis of individual rather than group performance (as was the case during this period) is a more effective productivity incentive; however, the Army College Fund and the enlistment bonus have positive market expansion effects, in addition to the skill-channeling effect, that need to be considered

Table 3.2 (continued)

Effect of Policy/Practice	Reference	Data	Notes
		Other Recruiting-Related Strategies	
Recruiters should emphasize job stability, job training, and educational opportunities provided by the military	Orvis and Gahart (1990)	1983 Survey of Military Applicants (n=6,857)	Recruiters are the main source of information for potential enlistees; results indicate that even after application, civilian job opportunities, social support for enlisting, college plans, and finances have large effects on enlistment decisions of young men; thus, emphasizing these aspects along with bonuses and educational benefits and (given the social support findings) directing the messages to family members and teachers would be very useful
Emphasizing the training aspects of the military might attract more enlistments	Borus and Kim (1985)	National Longitudinal Survey of Youth, 1979–1981 (n ≈ 12,600)	Among whites and younger black youth, desire for occupational training was a major motivation for entering the military; thus, more active promotion of the specific training opportunities and trying to match enlistees to desired jobs might help recruitment

Table 3.2 (continued)

Effect of Policy/Practice	Reference	Data	Notes
Those with negative intentions to enlist are an important source of enlistees	Kim (1982a, 1982b); Kim, Nestel, Phillips, and Borus (1980)	First and second round of interviews of the National Longitudinal Survey of Youth, 1979, 1980 (n ≈ 984 enlistees)	The most cited reason for enlistment in the military was training opportunities; black youth had particularly high intention to enlist; school attendance was the main reason for not enlisting among those who had talked to a recruiter but did not join; an analysis of job aspects indicated that armed forces personnel were less satisfied with the amenities, rewards, and motivation provided by their jobs than their civilian labor market counterparts; this was particularly so in the case of personnel in the ground arms (Army and Marine Corps) and may help explain differential recruiting appeal of the services
	Orvis, Gahart, Ludwig, and Schutz (1992); Orvis (1986)	11 Youth Attitude Tracking Study (YATS) administered between 1976 and 1981	About half of all enlistees come from the negative intention group, so focusing strictly on those with positive intentions is misguided; market research should identify factors that promote enlistment among persons at all intention levels; other findings: Stated enlistment intentions are significantly related to decision to enlist; aggregated intention information predicts regional enlistment rates; intentions data reveal more than demographics alone; and intentions to enlist under hypothetical options provide a useful addition to field test results

Table 3.3

Screening for Enlistment, Attrition, and Job Performance

Effect of Policy/Practice	Reference	Data	Notes
Screening for Enlistment and Attrition			
Raising AFQT requirements would screen out more blacks	Gorman and Thomas (1993)	Sample of young male high school graduates from the 1979–1987 National Longitudinal Survey of Youth (n=1,910)	AFQT scores used as proxies for general intellectual achievement; variation in intent to enlist depends on AFQT score, age, poverty status, and race; failing to account for general intellectual achievement tends to overstate the importance of race as a predictor of enlistment intentions; the highest intent to enlist is among those with low AFQT scores and those living in poverty; blacks tend to be overrepresented in both these groups
Attrition among women is related to work group relationships, family and career orientation, and stress; less important are recruiting, training, and placement	Royle (1985)	Surveys of a sample of 703 first-term Marine women and/or their supervisors from 1981 and 1982	To help decrease attrition among women, the Marine Corps should discourage women with the most traditional career and family goals from enlisting, help women develop coping skills, provide sex education, and improve work group climate as well as the climate toward women as a whole in the USMC
Women have higher attrition than men	Quester and Steadman (1990)	Female Marine recruits with four-year obligations accessed in FY81 to FY85	Attrition rates for first-term female Marines is 1.5 times higher than for males; long-term attrition rates are lower for females; models of first-term attrition yield similar results for males and females except for the pregnancy/parenthood variable

Table 3.3 (continued)

Effect of Policy/Practice	Reference	Data	Notes
High school graduation is the best predictor of first-term attrition; female recruits, especially in nontraditional MOSs, have higher attrition rates	Ross, Nogami, and Eaton (1984)	All female recruits and a 10 percent sample of male recruits entering the Army in 1976	Pregnancy and family-related reasons accounted for 37 percent of all female attrition
Expansion of the role of women could enhance combat readiness, if properly executed	Brown (1987)		It will be more desirable to have females of a higher mental category than lower-aptitude males; combat units should not be integrated because of the questions of cohesion and additional male strength required; services should not block specialty/ratings service-wide, rather exclude specific positions (as done in the Air Force); the public should be educated about possible female losses during war

Table 3.3 (continued)

Effect of Policy/Practice	Reference	Data	Notes
Women in the military have lower turnover rates than their civilian counterparts, suggesting that the military has been successful in their training and other human resource policies; no effect on turnover of being in traditionally female or traditionally male occupations in the military but among mixed occupations, changes towards a higher proportion female reduces turnover for women	Waite and Berryman (1985)	National Longitudinal Survey of Youth Labor Market Behavior, selected samples of males and females 16–21 years old employed in 1979 (n=1077, 1376)	Traditionally male and female occupations in the military were defined as those with 10 percent or less female and 75 percent or greater in the most comparable occupations in the civilian labor force; military exit rates were less than half of civilian exit rates; working unusual shifts or increased travel (both hypothesized to relate to job attractiveness) had no effect on turnover; black women in the services had substantially lower attrition than whites or Hispanics

Table 3.3 (continued)

Effect of Policy/Practice	Reference	Data	Notes
Small set of factors can reveal wide range of attrition risk among enlistees; suggests the usefulness of market segments as defined by seniors vs. graduates and education expectations (the latter works in opposite direction for seniors vs. graduates); this also helps explain why educational benefits are effective incentives; should enquire about employment history	Antel, Hosek, and Peterson (1987); Hosek, Antel, and Peterson (1989)	1979 DoD Survey of Personnel Entering the Military Service pooled with nonenlistee data from the National Longitudinal Survey of Labor Market Behavior Youth Survey; recruits tracked through 1984	Enlistment is negatively related to an individual's academic ability, education finances (more important for seniors), and employment opportunities (more important for graduates); attrition risk is related to senior/graduate status, positive/negative education expectations, stable/unstable civilian employment history, and short/long participation in DEP; additional factors include AFQT, especially for graduates, and months in labor force (graduates only); knowing the attrition potential of an enlistee cohort would be particularly useful when the opportunity cost of losing an enlistee is high (i.e., in situations marked by deteriorating enlistment/reenlistment conditions but stable or increasing force sizes)
Provides basis for recruit screening to reduce attrition; important factors in the models are race, gender, educational level, AFQT category, age, and term of enlistment	Zimmerman, Zimmerman, and King (1985)	Defense Manpower Data Center (DMDC) cohort data on Army accessions from FY79 to FY82 (n=358,000)	The most effective screening composite for males includes educational level, AFQT category, age, region, and race; for females the best composite includes term of enlistment, AFQT category, and race; these composites are designed only to predict attrition

Table 3.3 (continued)

Effect of Policy/Practice	Reference	Data	Notes
Probability of quitting the service is higher among nonhigh school graduates, those with dependents, younger recruits, and those with a history of troubles (legal, antisocial behavior, maladjustment); several suggestions offered to reduce quitting including more realistic portrayal of service life, counseling, and making personnel aware of true value of compensation	Stolzenberg and Winkler (1983)	Review of literature	Two concepts that measure how people evaluate their membership in their chosen group and the next best alternative—comparison level (CL) and comparison level for alternatives (CLalt)—are important in understanding why people quit; although compensation is very important, nonpecuniary factors may rival or exceed the influence of pay and benefits on decisions to quit; voluntary terminations could be reduced by (a) making personnel aware of the true value of their compensation, (b) using lump sum payments more extensively, (c) allowing some choice in the form in which compensation is paid, (d) providing mechanisms for resolving disputes, (e) counseling, and (f) realistic portrayal of military life
Along with educational attainment, preenlistment work history, and temporal variables (such as economic cycles, low scale hostilities, and cohort size) attrition is strongly influenced by institutional policy	Doering and Grissmer (1985)	Review of literature	An underlying assumption of attrition research has been that attrition rates will fall as the quality of cohorts improves; data from FY79–FY82 show that although the quality of enlistees rose dramatically, the attrition rates remained stable, suggesting that institutional policies may give rise to "creaming" of cohorts

Table 3.3 (continued)

Effect of Policy/Practice	Reference	Data	Notes
Recruiting high-quality recruits may add less to productivity than to cost	Congressional Budget Office (1986)	Projections of recruit quality, index of productivity, and costs, 1987–1991	The Army wanted to set recruiting objectives under which at least 90 percent would be high school graduates and 65–69 percent Cat. I–IIIA; higher-quality offers definite advantages: These recruits are more trainable, have lower attrition, thus reducing training and replacement costs, but increase recruitment costs; during the first term, high-aptitude soldiers perform better than lower-aptitude soldiers but groups of high-quality members do not always outperform groups with lesser ability; higher-aptitude soldiers tend to have lower reenlistment rates and may reduce the career force; although the force's overall productivity will increase with the percentage of high-quality enlistees in the force, long-run costs increase at a faster rate (although the analysis does employ a number of simplifying assumptions)
High school graduates have lower probability of first-term attrition, but also reenlist at a lower rate; minorities have lower attrition and higher reenlistment; reenlistment is related to civilian employment opportunities	Warner and Solon (1991)	1974 to 1983 accession cohorts into Army Infantry MOS 11	The first-term survival of graduates exceeds that of nongraduates by .15; however, graduates have only a slightly higher probability of continuing to a second term; blacks and Hispanics have a more than .05 higher probability of continuing to a second term; the elasticity of reenlistment probability with respect to military pay is about unity

Table 3.3 (continued)

Effect of Policy/Practice	Reference	Data	Notes
Recruit characteristics have an important effect on attrition, so initial screening can help reduce attrition	Buddin (1988)	1979 Survey of Personnel Entering Active Duty (AFEES) combined with active enlisted master and loss files	Work history of recruits has an important bearing on attrition: Persons with spells of unemployment or frequent job changes are more prone to attrition as are nonhigh school graduates, holders of General Equivalency Diplomas (GEDs), and older recruits; AFQT has a statistically significant but small effect on attrition; indicators of job match had no significant effect
Psychological screening may be important in reducing attrition	Mael and Ashforth (1995)	Survey of 2,535 male Army recruits in 1991	Earlier behavior and experiences as embodied in biodata may predispose one to identify with the Army (organizational identification or OID); in particular four biodata factors were found to be important: perceived congruence of personal interests and organizational activities; conformity to institutional expectations; preference for group attachments; and cognitively ambitious, achievement-oriented pursuits; both biodata and OID at the time of entry predicted subsequent attrition from 6–24 months; these constructs may help improve predictive validity, which could result in large savings

Table 3.3 (continued)

Effect of Policy/Practice	Reference	Data	Notes
Service practices and policies have an important effect on attrition	Buddin (1988)	High-quality nonprior service accessions, FY82–FY85	Cohort characteristics alone do not determine attrition rates; recruits of comparable quality have much different training attrition rates in some cohorts and at some bases in a given cohort; thus, although recruit characteristics can be used to rank prospective recruits by relative risk category, the interpretation and enforcement of service policies appear to critically affect the actual attrition level
First-term productivity growth is smallest for the most technical rating, suggesting that career mix should differ across broad categories of ratings; productivity differs by AFQT category and education	Marcus (1985)	Survey of first-term enlistees and supervisors, 1974 (n=6,558)	Supervisors were asked to rate a typical recruit's net productivity at several points in his first term; these were used to create average productivity values of a first-term recruit compared to a careerist; ordering of relative productivities is inversely related to the ordering of skill levels in the ratings, with the most technical ratings displaying the smallest first-term productivity; in addition, aptitude and education have a substantial effect on productivity

Table 3.3 (continued)

Effect of Policy/Practice	Reference	Data	Notes
AFQT scores have a significant effect on the performance of individuals and the performance of groups	Winkler, Fernandez, and Polich (1992)	Tests of 720 new Advanced Individual Training (AIT) graduates and 252 active duty personnel in MOS 31M, Multichannel Communications Equipment Operator	The study finds that the AFQT score has a direct, consistent effect on the ability of communications personnel to provide effective battlefield communications to Army units; the evidence suggests that AFQT scores have a sizable effect on group performance; groups that are on average "smarter" outperform other groups; the study concludes that a lowering of accession standards will substantially reduce the probability of operator success in operating and troubleshooting communications systems
Maintaining or improving the quality of Army recruits is cost-effective when productivity and readiness are considered	Toomepuu (1986)	Literature review and critique	Offers a more complete cost-effectiveness analysis than previous research by including better measures of cost-effectiveness and productivity; lowering the quality of manpower would reduce both combat and cost effectiveness
The effect of AFQT on first-term attrition varies by MOS, suggesting that a better matching of people to jobs might help reduce attrition	Manganaris and Schmitz (1985)	FY81 enlisted accession cohort; 13 MOSs with n>500 selected (n=40,776)	For similar individuals, the effect of AFQT on attrition varies by MOS; some MOSs display greater sensitivity to changes in AFQT scores; high school graduates have lower attrition than nongraduates but those holding GEDs are similar to those with diplomas; women have higher attrition than men; blacks have significantly lower attrition than either whites or Hispanics

Table 3.3 (continued)

Effect of Policy/Practice	Reference	Data	Notes
Education and aptitude are inversely related to attrition	Hawes (1990)	99 percent of all Marine enlisted accessions from 1983 to 1988	First-term attrition is analyzed using survival analysis and three predictor variables; education has the strongest effect, with high school graduates having the highest survival rate; aptitude level (AFQT score) is inversely related to attrition; presence of a moral waiver is associated with a slightly higher attrition rate
High school graduation is the best predictor of first term attrition; female recruits, especially in nontraditional MOSs have higher attrition rates	Ross, Nogami, and Eaton (1984)	All female recruits and a 10 percent sample of male recruits entering the Army in 1976	Pregnancy and family-related reasons accounted for 37 percent of all female attrition
Better screening of enlisted personnel could result in large savings	GAO (1997)	Attrition data on entry cohorts, 1986–1994	14 percent of new recruits leave during the first six months of service; main reasons are (a) inadequate screening for disqualifying medical conditions or previous drug abuse; (b) failure to perform adequately because of poor physical condition or lack of motivation; recommendations are to improve medical screening, move all drug testing to the Military Entrance Processing Station (MEPS), and strengthen recruiter incentives

Table 3.3 (continued)

Effect of Policy/Practice	Reference	Data	Notes
Better coding of attrition causes and timing would enhance efforts to better manage attrition-related policies	Laurence, Naughton, and Harris (1996)	Review of literature	Past attrition research has limited value for policy improvement; more detail is needed to understand different types of separations at different points in the first term of service; also, most attrition models lack the depth required to understand the causes for particular attrition decisions
Readily available background characteristics do not forecast the reason why an individual recruit may separate early; counseling or screening out individuals with social or emotional immaturity would help reduce attrition	Klein, Hawes-Dawson, and Martin (1991)	Entrants in FY79 or FY85 who left for an adverse reason (n=1,134)	Most recruits who leave early do so for a multiplicity of reasons: work/duty problems, training problems, minor offenses, mental health problems, and drug/alcohol problems are the most common; differences in individual recruit characteristics seemed unrelated to the reasons why a recruit was discharged; mental health problems were more likely to surface early in the term so counseling might help; screening for emotional instability would be useful but very expensive; exit interviews may prove helpful in understanding the reasons why recruits leave
Screening for Job Performance			
Most important characteristic for predicting performance is high school graduation	Quester and Olson (1988)	Center for Naval Analyses (CNA) database for all nonprior service Navy accessions between FY78 and FY86.	The most important recruit characteristic for predicting performance is high school graduation; graduates outperform GED holders and nongraduates in first-term attrition, retention beyond first term, desertion, demotion, and promotion

Table 3.3 (continued)

Effect of Policy/Practice	Reference	Data	Notes
AFQT scores are related to performance on Skill Qualification Test (SQT); high school graduation is related to retention through the first term	Fernandez and Garfinkle (1985)	Data on four Army jobs for the cohorts of FY77–FY80 and FY81	SQT scores and retention can be combined into a measure of performance based on months of qualified service (Qualified Man-Months—QMM); changing service-wide or even job standards can increase the numbers of QMM; in four jobs studied, the higher-quality cohort of 1981 had 13 to 91 percent more QMM than the misnormed cohorts of FY77–FY80; raising standards involves a tradeoff in the form of more costly recruits; it may be more effective to focus more on AFQT scores and less on high school graduation
Higher AFQT scores are associated with higher military performance; less training and experience are also needed for high AFQT recruits to perform well	Orvis, Childress, and Polich (1992)	Tests of 100 students and 200 unit personnel in the Army's MOS 24T (Patriot missile personnel) and personnel data including AFQT score	Tests involving simulated operation of Patriot missiles showed significant differences in performance by AFQT score; AFQT had a stronger effect than experience and length of individual training; unit training was the second most important factor in predicting performance
Pre-basic-training instruction in basic verbal and literacy skills increases completion of BMT, completion of the first term of service, and improves promotion probabilities for selected recruits	Thomlison (1996)	25,000 FY92–FY93 Navy recruits with low verbal expression scores on the AFQT; of these, 3,323 participated in Fundamental Applied Skills Training (FAST)	FAST participation is found to have sufficient benefits in performance and retention to suggest that all low-scoring recruits be placed in the course

Table 3.3 (continued)

Effect of Policy/Practice	Reference	Data	Notes
Technical school training is less expensive than training within a squadron (OJT); units and individuals have lower productivity in OJT settings	Smith (1986)	Interviews, observation, and cost data involving Air Force units at three bases where seven civil engineering specialty codes are found	This study compares the productivity of civil engineering airmen who were either trained in technical school or sent to direct-duty to be trained by active squadrons; the results of the research show that technical training school is more cost-effective; cost savings range from $18,784 to $43,508 per recruit for the seven specialties examined; cost savings are related to lost time and production within active units for both trainees and the trainers
Pre-accession training provided on a contract basis may be cost-effective	Beveridge (1991)	32 Air Force Specialty Codes were chosen and data gathered on training methods and types of attrition within each code	Currently, enlistees spend some time in DEP before reporting to BMT; after finishing the course, enlistees can go to a unit for OJT or to a military technical training center for specialized training; the proposal called for pre-accession technical training during DEP; this would allow enlistees to join the unit after BMT; the study found that pre-accession training could be cost-effective; however, the balk rate (dropping out before BMT) and subsistence package (allowance to be paid to enlistees during the training) effect the savings over the current policy; in addition, the quality and "blueness" (development of military qualities) of training are unknown; this may also hinder management control of the training production process (i.e., ability to respond to under- or overproduction of some Air Force Specialty Codes (AFSCs))

Table 3.3 (continued)

Effect of Policy/Practice	Reference	Data	Notes
The effect of AFQT on first-term attrition varies by MOS, suggesting that a better matching of people to jobs might help reduce attrition	Manganaris and Schmitz (1985)	FY81 enlisted accession cohort; 13 MOSs with n>500 selected (n=40,776)	For similar individuals, the effect of AFQT on attrition varies by MOS; some MOSs display greater sensitivity to changes in AFQT scores; high school graduates have lower attrition than nongraduates but GED holders are similar to diploma holders; women have higher attrition than men; blacks have significantly lower attrition than either whites or Hispanics
Confirms that ASVAB scores can be used to predict performance in initial training courses	Schaffer (1996)	CNA data on Marines in eight initial training courses in 1995 (n=7,314)	ASVAB composites are placing individuals with similar aptitudes into courses requiring those aptitudes; such effective job classification reduces costs by increasing productivity, reducing training school attrition, and eventually increasing reenlistment; however, the grading system used in many Marine training courses gives most students the same passing grade, making performance measurement and future creation of ASVAB composites difficult and expensive

Table 3.3 (continued)

Effect of Policy/Practice	Reference	Data	Notes
ASVAB predicts job performance; a new series of experimental tests measures non-cognitive, psychomotor, perceptual, and cognitive characteristics not now measured by the ASVAB; these could be very useful in screening individuals for occupations	Campbell and Zook (1996)	Second-tour NCOs; Longitudinal Validation I sample of 10,000 enlistees who entered the Army in 1986/87; the Longitudinal Second-Tour sample of 1,500 individuals in nine MOSs who had reenlisted and were two to three years into their second tour	The ASVAB is an excellent predictor of job performance for both first and second tour, specially in predicting technical task performance and leadership; the Experimental Predictor Battery developed in Project A measures certain aspects of personality and interests that contribute to a selection validity that is higher than for ASVAB alone; these tests can help provide guidance in selecting Army recruits and making optimal assignment decisions to match both the needs of the individual and those of the Army
Attrition rates can be projected before MOS assignment, which can be useful in designing allocation policies that reduce costs	Manganaris and Schmitz (1984)	DMDC cohort data for 1976–1978 on 76 Army MOS covering 90 percent of accessions	Three regression equations are developed to project the attrition rate of eight demographic groups to 76 MOSs; education, sex, and AFQT, along with MOS assignment are the independent variables; the rates generated by these equations show where important tradeoffs exist with respect to personnel allocation and the expected rate of attrition

Table 3.3 (continued)

Effect of Policy/Practice	Reference	Data	Notes
		Compensation-Related Issues	
The civilian wage index used for military pay adjustment—Employment Cost Index (ECI)—does not track recruit quality and retention outcomes well; a new index, called the Defense Employment Cost Index (DECI) may provide a better measure of the opportunity wage of current active duty personnel	Hosek, Peterson, VanWinkle, and Wang (1992)	Trend data in basic military pay disaggregated by age of workers, education, years of military service, pay grade, accession and retention data 1977–1989, trends in ECI and civilian pay	The ECI is a fixed base weight index intended to show the employment cost growth for a given bundle of labor where the weights represent the employment distribution by occupation/industry group, based on the most recent Census; although it is accurate, timely, and stable relative to changes in the age composition of the labor force and changes in the business cycle, it does not accurately reflect the opportunity wage of active duty personnel; the latter is much better measured by the DECI, which employs weights based on age, education, and occupation distribution of the military and then pairs the weights with their respective civilian wages; its advantages are accuracy, sensitivity to the age composition of the military and to cyclical fluctuations, and flexibility, because it can be computed for various subgroups; unlike the ECI, which actually showed a negative relationship between pay and accession quality and retention, the DECI had a positive correlation with both and could help guide the annual military pay adjustment process

Table 3.3 (continued)

Effect of Policy/Practice	Reference	Data	Notes
Intergrade pay skewness (larger pay differentials as grades increase) increases performance, incentive to supply effort at higher grades, and retention of most able individuals; may reduce cooperation and teamwork; intergrade differentials to elicit the same level of work effort and retention will be smaller, the larger the nonpecuniary reward differentials across grades	Asch and Warner (1994a)	Review of literature and application of a new, unified model applied to the military compensation system	Pay differentials must increase to provide individuals in higher grades with an incentive to provide effort; this is important because the marginal productivity of effort rises with grade; also, promotion rates decline in higher grades so larger differentials are needed to maintain a constant-effort incentive; if rewards are skewed too much, a kind of cutthroat competition may arise; thus, pay differentials will be narrower where the team aspect of production is greater; the privileges, power, and status associated with higher grades may reduce the intergrade differentials required to produce the same level of work effort and retention
Performance-based intragrade pay increases effort	Asch and Warner (1994a)		Effort is especially increased for individuals spending more time in a grade; unqualified climbing is also decreased

Table 3.3 (continued)

Effect of Policy/Practice	Reference	Data	Notes
The small number of distinctions in the pay table, salary compression among those distinctions, and current promotion policies work against optimal ability sorting and organizational efficiency	Rosen (1992)	Literature review	Finds that the current compensation and promotion policies are more suitable to an internal labor market with conscription
In-kind benefits increase costs and decrease efficiency	Asch and Warner (1994a)		In-kind benefits reward personnel for marriage and for having children and not for performance; personnel can best decide how to spend benefits to maximize efficiency
Compensation levels should take into account post-service earning potential by occupation to optimize the occupational, grade, and experience structure of the force	Stafford (1991)	Internal Revenue Service, Social Security Administration, and Army databases	Analysis of post-Army earnings suggests that compensation should be fine-tuned to avoid overpaying some occupations where civilian opportunities are less attractive; likewise, some occupations may be underpaid, thus encouraging shortened Army careers and loss of high-productivity years of service that balance high training costs

Table 3.3 (continued)

Effect of Policy/Practice	Reference	Data	Notes
Pay for human capital development would help improve productivity	Robbert, Keltner, Reynolds, Spranca, and Benjamin (1997)	Literature review, interviews	Such pay could reinforce desired behaviors such as high levels of fitness and marksmanship; one problem is that access to human capital development opportunities is uneven
Intragrade merit pay raises and individual performance bonuses may undermine intrinsic rewards and the shared culture of military service	Robbert, Keltner, Reynolds, Spranca, and Benjamin (1997)	Literature review, interviews	Interviews suggested that members were skeptical that the best performers can be identified in the short run; tolerance for greater use of such extrinsic rewards was low in core combat communities; bonuses might be better than merit pay raises: fewer administrative costs, more immediate effect, and in addition, the effect of a bad decision would be less enduring
Military pay is an important factor in meeting enlisted personnel quality and quantity goals; capping military pay below civilian pay would have negative effects on recruiting and retention	Hosek, Peterson, and Heilbrunn (1994)	DECI and ECI	Accession and retention are sensitive to pay levels when comparisons to civilian wages are made using the DECI; previous estimates of the gap between military and civilian pay levels have been based on the less-appropriate ECI and have supported a view of military pay as less important in enlistment and reenlistment decisions

Table 3.4

Sources of Entry

Effect of Policy/Practice	Reference	Data	Notes
Prior service accessions may be a useful alternative to nonprior service accessions	Fernandez and De Tray (1984)	FY74 nonprior service entry cohort tracked through 1981 (n=390,460)	Members of the FY74 cohort who became prior service accessions were representative of the cohort as a whole; they completed their first term more frequently than the average cohort member; in their second term, they did not complete their term or advance as rapidly as did reenlistees; need to consider carefully how expensive they would be to recruit and how promotion ladders would be affected
Some substitution of civilian personnel for military personnel may be possible in shore billets	Marcus (1985)		This would be particularly effective in ratings with low sea-to-shore rotation but would have the effect of increasing the sea-to-shore rotation of personnel in these jobs

Table 3.4 (continued)

Effect of Policy/Practice	Reference	Data	Notes
Policies prohibiting lateral entry require constant thinning of the force and distort compensation levels	Asch and Warner (1994a)		Limits on lateral entry require the military to "overstock," i.e., to hire a large new-entrant pool to ensure a sufficient supply of qualified individuals to the higher ranks; the cost of overstocking is the firm-specific investment a hierarchical organization must make to grow its upper-level employees; in addition, the organization must generate some turnover at the higher levels to provide promotion opportunities for lower-level individuals; thus, junior workers are overpaid relative to their outside opportunities but equal to their higher internal productivity; to ensure retention, senior workers are overpaid relative to their outside opportunities but underpaid relative to their internal productivity
Relaxed lateral entry from the reserve may provide some flexibility; lateral entry from the civilian sector may cause problems	Robbert, Keltner, Reynolds, Spranca, and Benjamin (1997)	Literature review, interviews	Lateral entry may prove useful in technical skills but this needs to be considered carefully; reservists would more easily be absorbed into the military but the range of technical skills acquired would be limited to those found among reservists; hiring civilians into middle or higher grade positions may be more problematic: They lack general military skills, they may not have the same respect for authority, and such a policy may be viewed as inequitable by other members

Table 3.5

Developing and Training

Effect of Policy/Practice	Reference	Data	Notes
Increased use of civilian training is feasible, but must be done after further evaluation to determine how and when it will be most effective	Hanser, Davidson, and Stasz (1991)	General review of military and civilian training environment	Main conclusions are that (1) many military occupations are amenable to civilian training, (2) former and existing programs have not been adequately evaluated, (3) civilian-provided Initial Skill Training (IST) appears to have benefits in some circumstances, and (4) there are institutional barriers to implementation
Technical school training is less expensive than training within a squadron (OJT); units and individuals have lower productivity in OJT settings	Smith (1986)	Interviews, observation, and cost data involving Air Force units at three bases where seven civil engineering specialty codes are found	This study compares the productivity of civil engineering airmen who were either trained in technical school or sent to direct-duty to be trained by active squadrons; the results of the research show that technical training school is more cost-effective; cost savings range from $18,784 to $43,508 per recruit for the seven specialties examined; cost savings are related to lost time and production within active units for both trainees and the trainers

Table 3.5 (continued)

Effect of Policy/Practice	Reference	Data	Notes
Limited MOS consolidation and crosstraining can reduce costs and increase organizational flexibility without significant readiness erosion; significant shift from AIT to OJT produces problems for readiness and training quality for complex and technical skills	Wild and Orvis (1993)	Selected sample of Army helicopter maintainers and maintenance supervisor; data included surveys, Enlisted Master File data, and field maintenance data	Study focused on consolidation of 13 MOSs involving helicopter maintenance; reduction to five or fewer MOSs is likely to cause significant decrements to unit capability and depth of skills and experience; reduction to more than five MOSs provides the cost and flexibility benefits desired while minimizing negative effects; increased use of on-the-job training causes problems for readiness and limits opportunities for broad training experience
Increased use of distributed training, Training Aids, Devices, Simulators, Simulations (TADSS), and civilian training may be most cost-effective when used for MOSs that are grouped by dominant tasks, cost to train, ability requirements, and civilian exchange-ability	Winkler, Kirin, and Uebersax (1992)	Army doctrinal publications, published literature, DoD and civilian job training data sources	Training-related characteristics of Army occupations are used to develop dimensions that may be used to develop effective training methods; for example, civilian training and experience may be most cost-effective for MOSs that are high in training costs, high in civilian exchangeability, and low in ability requirements

Table 3.5 (continued)

Effect of Policy/Practice	Reference	Data	Notes
Training costs can be reduced by consolidating training programs across courses, locations, and military occupations; such consolidations must be chosen carefully to minimize possible tradeoffs in individual and unit capability; other changes that can reduce costs in certain specific circumstances include increased use of computer-based training, simulation technology, distributed learning, and civilian training	Winkler and Steinberg (1997)	Previous RAND research published in over a dozen reports	Summarizes results and insights from RAND studies, mostly from the Arroyo Center; these studies show that there are a number of ways to make training less costly and more efficient; any restructuring must be done on a case-by-case basis with new methods matched to specific circumstances; it is not possible to say that civilian training should be increased across-the-board or that a major consolidation of MOS is needed; what is needed is ongoing systematic, quantitative assessment of education and training innovations to establish effectiveness and appropriateness in different military settings

Table 3.5 (continued)

Effect of Policy/Practice	Reference	Data	Notes
Training resources may be conserved when jobs are analyzed and matched to programs of instruction that have been subjected to resource and cost analyses	Farris, Spencer, Winkler, and Kahan (1993)	Surveys, performance measures, job attributes, and current training methods were collected for Army Cannon Fire Direction Specialists	This research develops improved techniques for identifying alternative approaches for conducting individual training and analyzing their cost implications; for cannon fire specialists the length of the course of training may be shortened and a substantial number of tasks may be taught using computer-based technology; these changes would provide significant cost savings while still meeting fundamental training objectives
Large-scale training exercises will likely need to be replaced in large part by simulations because of increasing cost and political and environmental constraints	Allen (1992)	Results of experiments using differing training methods during Army exercises in Germany in 1989 and 1990 (REFORGER exercises)	While training by simulation was found to be less expensive and in some cases more effective, there are areas where current simulation techniques need significant improvement; effective simulations require better representation of different combinations of forces, more varied types of battles, inclusion of "bad" information, and better representation of the "fog and friction" of war
Gender-neutral occupational performance standards have not been carefully evaluated	GAO (1996a)	Review of gender-neutral occupational performance standards in the military services	The services have few data on implementation of gender-neutral performance standards and screening of service members to ensure that they can meet the physical demands of their occupations; assessments of successful implementation have been based largely on a lack of complaints

Table 3.5 (continued)

Effect of Policy/Practice	Reference	Data	Notes
Skill training rates may be reduced by allowing fewer waivers of ASVAB prerequisites, particularly in the most difficult training programs	Yardley (1990)	Enlisted Training and Tracking File data for 1988 for the 13,358 Navy enlisted personnel enrolled in the 15 training pipelines with the highest overall attrition rates	Academic attrition during skill training occurs at a higher rate for personnel who failed to meet ASVAB prerequisite scores but were enrolled with a waiver; academic failure rates of waivered students varies considerably across rating pipelines
Reducing the number of PCS moves would reduce personnel readiness	Thompson, Krass, and Liang (1991)	Navy Enlisted Personnel Management Center data	Over a five-month period, a reduction from 17,300 to 14,495 moves would reduce costs but would result in a drop in readiness from 1.85 to 2.30 on a measure developed in this study
Personnel practices characterized as quality management were instituted including job series consolidation, pay banding, gainsharing, elimination of individual performance appraisal, a focus on teams, on-call hiring authority, and changes in ways to determine supervisory grade and	Gilbert (1991)	General review of the first three years of the PACER SHARE demonstration project, covering over 1,800 federal civil service employees in the Sacramento Air Logistics Center; four other logistics centers serve as comparison groups	This review found slightly more positive results than did the later three-year review conducted by RAND (Orvis, Hosek, and Mattock, 1993); the focus here is the use of the quality approach to management as found in the Total Quality Management (TQM) literature and also on the use of incentives and the team basis of organization; some findings: a 30 percent decrease in number of civilians and supervisors; reduction in labor/management grievances; payroll savings of $7 million; multiskilled work force; national recognition for quality; improved quality of work life; and improved employee respect for their leadership on some dimensions

Table 3.5 (continued)

Effect of Policy/Practice	Reference	Data	Notes
pay; reported improvements were reductions in employees and supervisors, better labor relations, reduced costs, more skilled workforce, and improved quality of work life			
Innovative personnel practices were instituted including job series consolidation, pay banding, gainsharing, elimination of individual performance appraisal, on-call hiring authority, and changes in methods for determining supervisory grade and pay; results included increased flexibility, job satisfaction, trust, retention, and organizational attachment; work quality, timeliness, and overall labor cost showed no significant improvement	Orvis, Hosek, and Mattock (1993)	Data from the first three years of the PACER SHARE demonstration project, covering over 1,800 federal civil service employees in the Sacramento Air Logistics Center; four other logistics centers serve as comparison groups	The personnel practices included in this experiment were implemented as a group, making evaluation of the effects of any one policy problematic; while significant cost savings were not found, there were enough other positive effects to make the results encouraging; assessment after a longer period of implementation may show improvements that are expected to be more long-term in nature; there are benefits associated with more broadly trained employees, higher retention, better morale, and increased flexibility to meet varying workloads

Table 3.6

Promotion Policies

Effect of Policy/Practice	Reference	Data	Notes
Moving 10 percent of the Army's NCO content down to more junior grades will reduce costs by $146 million to $169 million; proposed change is from 49.9 percent to 45 percent of the enlisted force endstrength	Orvis and Way-Smith (unpublished)	POF (Programmed Objective Force) estimates, ODCSPER data, TRADOC Resource Factor Handbook, and other Army manpower data sources	Savings will not begin until the fourth year of the four year phase-in period due to offsetting costs associated with accessions, training, and separation; savings will vary depending on whether the shift will be to E1–E4 or to E4 only; effects on readiness are not assessed in this study
Up-or-out policies increase exit rate of unqualified and mismatched workers and decrease costs of monitoring individual performance	Asch and Warner (1994a)		Up-or-out policies provide a bureaucratic way of separating poor performers in a setting where effective, continuous monitoring is difficult and supervisors have little incentive to demote them even when recognized
Promotion tempo (time to E5) affects reenlistment for second term	Buddin, Levy, Hanley, and Waldman (1992)	DMDC master files for Army and Air Force personnel finishing four-year first terms in FY83–FY89	A 10 percent slowdown on promotion tempo relates to 14 and 8 percent decreases in Army and Air Force retention rates. Previous studies testing the retention effects of pay AFQT score, graduation, and occupation are biased by the omission of promotion tempo

Table 3.6 (continued)

Effect of Policy/Practice	Reference	Data	Notes
Reducing the weight of occupational differences in promotion selections may have mixed effects	Robbert, Keltner, Reynolds, Spranca, and Benjamin (1997)	Literature review, interviews	If promotion is viewed as a reward mechanism, reducing the weight of occupational differences in promotion may help productivity and equity but at the same time, it reduces the capacity of the services to favor core functions/occupations; promotions based on occupation are often seen as unearned and not meeting the equity standard of fairness
Exceeding career content limits increased cost by $73.9 million during the FY86 through FY89 period	GAO (1991)	Air Force and Army data for FY86 through FY89	While the Air Force and Army complied with most DoD enlisted force management requirements, both exceeded the POF target for career content (enlisted career personnel with more than four years of service); a more senior force results in increased personnel costs in terms of increased military pay and retirement benefits; in addition, not managing within targets creates cycles of peaks and valleys as the services are forced to bring in fewer recruits to stay within authorized end-strength limits; DoD should establish more explicit criteria for identifying the levels of seniority needed; enlisted seniority has been on an upward trend throughout the 1980s

Table 3.7

Retention Policies

Effect of Policy/Practice	Reference	Data	Notes
		Family Support	
Spouse support programs including helping spouses find work at new duty locations, helping resolve incompatibilities in work schedules, would have a positive effect on retention	Lewis (1985)	Subsample of enlisted personnel and spouses in the Air Force Family Survey (AFFS), 1984 (n=540)	Spouses generally viewed Air Force life as more stressful than civilian life, yet were supportive of and involved in the Air Force; major sources of stress were disruptions due to work schedules, TDYs, reduced employment opportunities due to transfers; career intent and job satisfaction of enlisted personnel were closely linked to spouse attitudes and other family variables
Financial incentives, job changes, or trips home may help personnel to extend tours in Europe thus reducing PCS costs	Ozkaptan, Sanders, and Holz (1984)	Surveys of 1,000 Army families in USAREUR	About half of those surveyed were undecided about extending and represent a target of opportunity; family and job-related reasons were the most important factors for the indecision; thus, bonuses, trips home, and/or job changes may help increase the likelihood of extensions; family support programs are more important than improvement of community support services alone

Table 3.7 (continued)

Effect of Policy/Practice	Reference	Data	Notes
Job satisfaction and satisfaction with family life (particularly spousal attitudes) are the most important determinants in the decision to extend European tours; programs to improve family life and support and job satisfaction might be helpful	Lakhani, Thomas, Anderson, Gilroy, and Capps (1985)	The Army Families in Europe Survey, 1983 (n=282)	Spousal perception of family satisfaction was particularly important in the model so orientation programs and other support programs might be valuable in helping them adjust; older, more experienced servicemen were less likely to extend
Retention of high performing soldiers is related to opportunity for advancement and quality of life issues for families	Rakoff, Griffith, and Zarkin (1994)	Survey of 5,299 Army, male, enlisted personnel merged with loss data and supervisor ratings	Retention intentions for high-performing soldiers are related to perceptions of opportunity in military life for advancement, service to country, and excitement; also important were the quality of place for children and career opportunities for spouse while in the military as compared to civilian life
Spouse attitudes towards retention are based on a number of factors	Rosenberg, Vuozzo (1989)	1987 Annual Survey of Army Families (n=12,000)	Spouse attitude toward retention is related to the soldier's satisfaction with the job, security and stability of the job, and retirement pay and benefits; benefits, rather than pay, represent the single most salient financial issue. Respondents are concerned with a perceived erosion of benefits such as medical and retirement

Table 3.7 (continued)

Effect of Policy/Practice	Reference	Data	Notes
Presence and use of on-site child development center does not appear to affect the incidence of child-care related work interference among military personnel nor the probability that child care experiences will influence career decisions	Lofink (1990)	Survey of Navy personnel at eight commands who were identified as having a dependent under 13 years of age (n ≈ 690)	Usage rate of military child care centers was quite low (partly due to lack of space, and partly due to inconvenience or dissatisfaction with hours or quality); incidence of work interference was higher for those with preschool children, in families with full-time working spouses, single women parents, married enlisted with some college education, and for personnel assigned to installations without on-site child care centers; however, the presence or absence of on-site centers did not significantly increase the probability of leaving because of child care experiences
		Reenlistment Incentives	
SRBs are shown to be cost-effective for filling specialties with low reenlistment and high costs; bonuses should be used continuously in these specialties and should target personnel in Zone A (3–6 years); retention effects are large and persistent	Walker (1991)	Air Force Enriched Airman Gain/Loss file	This report describes the design of the Air Force's Enlisted Force Management System (EFMS) and gives examples of how the model can be applied to personnel management policy questions

Table 3.7 (continued)

Effect of Policy/Practice	Reference	Data	Notes
Selective Reenlistment Bonuses have a very large effect on the term of enlistment (TOE) chosen by Air Force reenlistees	Carter and Hackett (1988)	30 percent random sample of airmen facing second- and third-term decisions between 1979 and 1984	Bonuses increased by length of term chosen; among second-termers, 90 percent of persons not offered a bonus selected minimal four-year terms while fewer than a third offered a bonus selected this short term; among third-termers, bonus eligibility had a strong effect but was somewhat less important among the oldest NCOs
Retention rates in technical MOSs can be raised by increasing the SRB; the SRB should vary by major Army occupational groups (Infantry, Technical, and Support)	Haber, Lamas, and Eargle (1984)	Data on Army personnel, continuation rates, recruitment and training costs, and SRB payments by major occupational group, FY81–FY83	Results for E-4s show that the actual SRB for combat arms is almost equal to the SRB calculated by the model but is set much lower than would be optimal for technical occupations and much higher for support services; retention in technical MOSs could be increased by raising the SRB; this could be paid for by reducing the SRB for support services
Pay elasticities vary widely across occupational categories, suggesting that occupation-specific pay elasticity should be used to determine bonus increases for ratings with manning shortages	Goldberg and Warner (1982)	First-term and second-term retention, FY74–FY80 (n=261,904 and 66,553, respectively)	The study examined reenlistments (for three years or more) and extensions (for less than three years); rates are highly sensitive to military pay but also differ across occupational categories suggesting that an all-Navy pay coefficient would give misleading results if applied to ratings with unusually high or low pay responsiveness

Table 3.7 (continued)

Effect of Policy/Practice	Reference	Data	Notes
Reenlistment bonuses are a powerful tool for controlling retention in targeted occupations and are effective countercyclical tools	Hosek and Peterson (1985)	Retention data for FY76–FY81 by occupation, supplemented with economic variables	Higher reenlistment bonuses increase both the rate of retention but also, more importantly, the expected manyears of active duty service in a given occupation because members are willing to commit for longer terms; while both higher military pay and bonuses increase retention, bonuses produce more expected manyears and may be more cost-effective in offsetting the detrimental effects of lower unemployment rates
Adherence to improved procedural basis for allocating Selective Reenlistment Bonuses would be more cost-effective and make the SRB budget more defensible	Fernandez (1989)	Review of SRB program	A model for allocating Selective Reenlistment Bonuses is developed that minimizes costs and maintains flows into specialties that are critical to the defense mission
Retention bonuses (paid under the SRB program) should be awarded only for reenlistments in skill categories that are in short supply	GAO (1995a)	Number and cost of reenlistment bonuses and separation bonuses, fill rates for skills receiving	Services are providing SRBs to skills with high fill-rates (90 percent or more) and in which many higher-skill-level members were being paid to leave the service; DoD defended the management of its SRB program, stating the SRBs and separation incentives were aimed at different segments of the force and that high fill-rates did not translate into adequate manning of those skills

Table 3.7 (continued)

Effect of Policy/Practice	Reference	Data	Notes
A one-level increase in the SRB is estimated to increase first-term infantry reenlistment rates by 2.2 percent; second-term rates, by 1.7 percent	Smith, Sylvester, and Villa in Gilroy, Horne, and Smith (eds.) (1991)	A 25 percent sample of all Army enlisted personnel who entered military service from FY74 through FY84	Soldiers serving in technical occupations in the Army are more sensitive to changes in basic pay levels when making reenlistment decisions; pay elasticities were 1.9 for mechanical maintenance specialties and 1.3 for infantry specialties
Lump-sum bonuses are more cost-effective than installment bonuses at the first-term retention point	Hosek and Peterson (1985)	Retention data for FY76–FY81 by occupation, supplemented with economic variables	On a cost-equivalent basis, lump-sum bonuses produce higher reenlistment and retention rates and lower extension rates; thus, lump-sum bonuses cause more people to shift to a longer contractual term; the advantage of lump-sum bonuses increases with increasing bonus amounts
Higher extension bonuses would increase probability of extending European tours of duty; lump-sum bonuses were more cost-effective than higher installment bonuses	Lakhani, Thomas, Anderson, Gilroy, and Capps (1985)	The Army Families in Europe Survey, 1983 (n=282)	Both installment and lump-sum bonuses were cost-effective relative to PCS costs; lump-sum bonuses increased the probability of extension more than installment bonuses; higher bonuses should be paid to older servicemen and those assigned to combat units

Table 3.7 (continued)

Effect of Policy/Practice	Reference	Data	Notes
Other Factors Affecting Retention			
High school graduates have lower probability of first-term attrition, but also reenlist at a lower rate; minorities have lower attrition and higher reenlistment; reenlistment is related to civilian employment opportunities	Warner and Solon (1991)	1974 to 1983 accession cohorts into Army Infantry MOS 11	The first-term survival of graduates exceeds that of nongraduates by .15; when attrition is taken into account graduates have only a slightly higher probability of continuing to a second term; blacks and Hispanics have a more than .05 higher probability of continuing to a second term; the elasticity of reenlistment probability with respect to military pay is about unity
AFQT and education are useful as enlistment standards but other unobserved ability factors as measured by rank achieved in first term and the speed with which that rank is achieved should be given greater weight in reenlistment decisions; measured by this quality index, the military is	Ward and Tan (1985)	Nonprior service male FY74 entry cohort followed through 1981	The study estimates indexes of quality for each of eight military occupations that consist of entry-level characteristics like education and AFQT and a catch-all collection of other factors called "unobserved ability"; AFQT scores have a small quantitative effect on the quality index compared to education, whereas the unobserved component that reflects military-specific ability explains a good deal of the variation in job performance; the conclusion is that (except for the Air Force, which has relatively homogeneous entry cohorts), entry-level characteristics are relatively unimportant as a predictor of first-term quality; rank achieved at the end of the first term and the speed with

Table 3.7 (continued)

Effect of Policy/Practice	Reference	Data	Notes
successful in retaining those with higher military-specific ability both during the first term and at ETS			which that rank is achieved should receive larger weight in reenlistment decisions than entry-level characteristics
Unemployment has a positive effect on enlisted retention but pay elasticities are much larger than unemployment elasticities, so decreases in unemployment could be offset by small changes in military pay	Goldberg (1985)	First-term reenlistment decisions, FY77–FY84, for 72 ratings	For some of the ratings, unemployment had a larger effect on the extension rate than on the reenlistment rate; the pay elasticities (with respect to Regular Military Compensation—RMC) were about three to five times larger than the unemployment elasticities; thus, a 10 percent decrease in the unemployment rate during this time period could have been offset by a 2–4 percent increase in military pay; because sea pay is omitted from the model, these elasticities may be biased upward
The effect of a raise in base pay on reenlistment rates varies by MOS group, with differences believed to be related to similarities between military and civilian jobs	Smith, Sylvester, and Villa in Gilroy, Horne, and Smith (eds.) (1991)	Panel data on a 25 percent sample of recruits who were accessed between FY74 and FY84 and entered infantry, mechanical maintenance, or administration careers	The elasticity of the reenlistment rate with respect to an increase in base pay is 1.3 for first-term infantry soldiers it is higher for mechanical maintenance (1.8) and for administration (1.9)

Table 3.7 (continued)

Effect of Policy/Practice	Reference	Data	Notes
Sensitivity of reenlistment decisions to pay vary across quality groups; the most able personnel are most sensitive to changes in pay, bonuses, and the economy; advancement to E-5 is very important in the reenlistment decision; career sea duty pay appears to increase voluntary extensions at sea	Marcus (1985)	Survey data from 1974 survey matched to the Enlisted Master Record and followed forward until reenlistment or separation (n=6,558)	High school graduates and whites are less likely to reenlist; the effect of AFQT was insignificant; increases in the SRB had a large effect as did unemployment, and these differed widely across mental groups with the largest effects present for the top groups; the effect of pay grade was quite dramatic, suggesting that advancement to E-5 in itself may be an effective retention tool
Intention to stay and perceived ease of finding another job are significantly related to decision to stay	Cooper (1991)	Survey of Air Force enlistees within one year of ETS in selected Air Force Specialty Codes (AFSCs), 1984–1986 (n=414)	Regional unemployment rate or tenure did not directly affect turnover; neither regional unemployment nor cognitive ability were related to perceived ease of movement; intention to stay was closely correlated with turnover as was perceived ease of movement; this was a relatively homogeneous sample, which may have attenuated some of the relationships

Table 3.7 (continued)

Effect of Policy/Practice	Reference	Data	Notes
	Steel (1996)	Sample of 402 enlisted personnel at Randolph Air Force Base	Reenlistment decisions are in some part dependent on labor market conditions for skill areas and also on how personnel perceive their marketability and range of choices
	Rearden (1988)	1985 DoD Survey of Officer and Enlisted Personnel merged with loss files sample of 6,328 Navy, male, enlisted personnel within 12 months of a reenlistment decision	Intention to reenlist is strongest predictor of actual reenlistment; age, marital status, pay grade, and satisfaction with military life also influence reenlistment behavior
Job satisfaction is important in the decision to remain in the service, as were nontraditional attitudes among females	Kim (1982b)	Second round of interviews of the National Longitudinal Survey of Youth, 1980 (n ≈ 984 enlistees)	Other important factors are marital status and presence of dependents, both of which have positive effects on reenlistment

Table 3.7 (continued)

Effect of Policy/Practice	Reference	Data	Notes
Retention of high-performing soldiers is related to opportunity for advancement and quality of life issues for families	Rakoff, Griffith, and Zarkin (1994)	Survey of 5,299 Army, male, enlisted personnel merged with loss data and supervisor ratings	Retention intentions for high-performing soldiers are related to perceptions of opportunity in military life for advancement, service to country, and excitement; also important were the quality of place for children and career opportunities for spouse while in the military as compared to civilian life
Promotion tempo (time to E-6) affects reenlistment for second term	Buddin, Levy, Hanley, and Waldman (1992)	DMDC master files for Army and Air Force personnel finishing four-year first terms in FY83–FY89	A 10 percent slowdown on promotion tempo relates to 14 and 8 percent decreases in Army and Air Force retention rates; previous studies testing the retention effects of pay, AFQT score, graduation, and occupation are biased by the omission of promotion tempo

Table 3.8

Separating and Retirement Policies

Effect of Policy/Practice	Reference	Data	Notes
Hybrid plans containing both a lump sum and an annuity may prove flexibility needed to separate personnel in different YOS groups in the event of downsizing	Grissmer, Eisenman, and Taylor (1995)	Model of voluntary separation offers and acceptance rates	The study evaluated the DoD's voluntary separation payment plans designed to reduce force size; separation offers need to be evaluated using a range of criteria including cost, efficiency, and equitable treatment of those staying and those leaving; an important question is how to target offers within YOS groups so as to preserve experience and high quality; hybrid plans can satisfy equity concerns at both ends of the YOS spectrum: for lower YOS, a lump-sum payment equal to separation pay and a deferred annuity at age 65; for higher YOS, the plan might offer a lump sum plus an annuity starting at an earlier age; hybrid plans also would ameliorate the long-term regret problem (since everyone would have a long-term annuity) and the lump-sum portion could be tailored to achieve desired acceptance rates
Retired pay gives strong effort incentive to mid-level personnel, but less incentive for younger and older workers	Asch and Warner (1994a)		The retirement system decreases needed mid-career separations; effort of younger personnel is reduced along with promotion opportunities

Table 3.8 (continued)

Effect of Policy/Practice	Reference	Data	Notes
A system that postpones annuities until age 60, gives separation benefits and vesting beginning at 10 YOS, some increase in active duty pay, and increased skewness of the pay table would maintain the force structure and cost levels of the current system while increasing productivity (individual effort, retention and advancement of high-ability individuals, and flexibility in personnel policies)	Asch and Warner (1994b)	Empirical application of military compensation system model developed in Asch and Warner (1994a)	The use of variable cash separation payments becomes the force management tool in this proposed system; variations in the size of payments and eligibility requirements can control the grade-by-experience distribution as well as flows through different skill areas.
To reduce costs, increase management flexibility, and maintain personnel trust, the retirement system should retain 20-year vesting, delay benefits to age 55, give partial benefits at 10 YOS, and continue noncontributory system.	Patten (1986)	Review of U.S. military retirement system and those of other nations and some private firms	Comparisons are made to Canada, the Federal Republic of Germany, the United Kingdom, and three large U.S. firms

Table 3.8 (continued)

Effect of Policy/Practice	Reference	Data	Notes
An analysis of the 1986 Military Retirement Reform Act reveals possible negative side effects, including larger-than-expected losses of personnel, a decrease in average length of service of over 10 percent, and retention decreases that are highest among higher-quality personnel	Walker (1991)	Air Force Enriched Airman Gain/Loss file	This report describes the design of the Air Force's EFMS and gives examples of how the model can be applied to personnel management policy questions
Increasing the portability of deferred benefits may be needed to attract lateral entry but would likely increase loss rates at all points between the new and current vesting points	Robbert, Keltner, Reynolds, Spranca, and Benjamin (1997)	Literature review, interviews	This would align military human resource management more closely with that of the civilian sector where early vesting is required by law

Chapter Four

CONCLUSIONS

The world is changing rapidly and in ways that will have a significant effect on enlisted force requirements and management of the force. The military services will need greater organizational effectiveness, flexibility, and adaptability. It is not entirely clear that in such a changing environment, past policies and practices will be entirely successful in meeting organizational objectives, nor is it clear that lessons we have learned from the past will apply in the future. Nonetheless, manpower planners need information on whether and how these policies and practices have worked, so that they may better understand how to change them or when to seek new and innovative practices. The objective of this report is to fill that information gap.

This review has shown the effectiveness of many current service policies and practices, when considered individually; what is important for the future, however, is to integrate these policies and practices to improve organizational performance. In this respect, the strategic human resource management approach adopted by the 8th QRMC holds promise for the future (DoD, forthcoming).

ANNOTATED BIBLIOGRAPHY

Author   Allen, Patrick D.

Title   *Simulation Support of Large-Scale Exercises:*
            A REFORGER Case Study

Report#   RAND/R-4156-A

PubDate   1992

PubData   Santa Monica, CA: RAND

Subject   training, cost, performance, Army, military, development

Abstract

This report describes an analysis of the Caravan Guard (CG) 89 and Centurion Shield (CS) 90 exercises. The study examines four different exercise training modes (both live and simulated) employed in CG 89 and CS 90 exercises: field training exercise, command field exercise, command post exercise; and computer-assisted exercise. The analysis leads to three recommendations for future large-scale multi-echelon exercises. First, exercises should consist of a single training mode and that should be simulation. Second, if simulations become the primary mode, a number of limitations affecting the current family of simulations must be overcome. Broad areas needing improvement include the representation of the effect of combined arms, the types of battles, aspects of how the operational level of war is depicted, the "fog and friction of war," and intelligence functions and products. Third, whenever possible, exercises should include both active and reserve component units and forces and other services and nations.

Author	Antel, John, James R. Hosek, and Christine E. Peterson
Title	*Military Enlistment and Attrition: An Analysis of Decision Reversal*
Report#	RAND/R-3510-FMP
PubDate	1987
PubData	Santa Monica, CA: RAND
Subject	accession, attrition, recruiting, personnel flow, military, background characteristics, recruit screening

Abstract

This report presents a theoretical discussion and empirical analysis of enlistment and first-term attrition. The theoretical discussion gives rise to hypotheses about enlistment and attrition. The enlistment hypotheses take a supply view, treating military service as an alternative to further schooling or to work. The attrition hypotheses are inherently two-sided, considering first the value of enlistment to the individual and the likelihood that he is more prone to disappointment due to poor planning, and second the value of the individual to the service and the chance that the service's eligibility screens were unable to identify low-productivity prospects. The empirical analysis is directed to the two prime recruiting markets from which the services draw high-quality male enlistees: high school seniors and nonstudent high school graduates. The study estimates sequential probit models for seniors and graduates separately, for both enlistment and six-month attrition and enlistment and 35-month attrition. The model produces estimates of the effect of individual characteristics on enlistment and on attrition, and controls for unobserved factors affecting both outcomes. The findings suggest that a small set of factors can reveal a wide range of attrition risk among enlistees. The factors are senior/graduate status, positive/negative education expectations, stable/unstable civilian employment history, and short/long participation in the Delayed Entry Program.

Author	Arguden, R. Yilmaz
Title	*Personnel Management in the Military: Effects of Retirement Policies on the Retention of Personnel*

Report#   RAND/R-3342-AF

PubDate   1986

PubData   Santa Monica, CA: RAND

Subject   retirement, transitioning, cost, military, personnel flow, compensation

Abstract

Many studies of the military retirement system are based on models whose structures are likely to be changed by the policy interventions that they analyze. Such models could lead to seriously biased predictions of the retention effects of alternative retirement systems. This report examines the adequacy of the existing retention models for retirement policy analysis, quantifies their limitations, suggests improvements, and develops a simulation methodology to test the suggested and future improvements. It also examines the importance of paying analytical attention to the inputs of the retention models.

Author   Armstrong, Bruce, and S. Craig Moore

Title   *Air Force Manpower, Personnel, and Training: Roles and Interactions*

Report#   RAND/R-2429-AF

PubDate   1980

PubData   Santa Monica, CA: RAND

Subject   personnel flow, accession, development, Air Force, military

Abstract

Provides the first consolidated summary of the Air Force's Manpower, Personnel, and Training (MPT) system, and describes the formal and informal functions of the system's three components. The components obviously interact: "Manpower" determines requirements for people and distributes budget-approved authorizations; "Personnel" determines management policies and tries to fill authorized positions with the right people; and "Training" recruits, classifies, and trains enlisted personnel. The report emphasizes links among the components and explains how the system relates to the

Planning, Programming, and Budgeting System. The Air Force can therefore use the report to identify needed improvements and to introduce newcomers to the MPT system's structure and functions.

Author	Asch, Beth J.
Title	Designing Military Pay: Contributions and Implications of the Economics Literature
Report#	RAND/MR-161-FMP
PubDate	1993
PubData	Santa Monica, CA: RAND
Subject	compensation, military, retirement, personnel flow, transitioning, promotion, development, pay table

Abstract

What should be the structure of military compensation for active-duty personnel? This broad question is the focus of the review of the economic literature presented in this report. The review addresses a more specific question, How should military basic pay be designed? Some of the key guidelines derived from the survey are that (1) individuals in occupations or positions with disamenities (e.g., greater injury/death/health risks) must receive higher pay than those in occupations with amenities; (2) compensation should rise with grade or with hierarchical level; (3) the intergrade compensation spread should increase with grade; (4) promotion policy can increase each individual's motivation and performance; (5) explicit up-or-out policies can sometimes be replaced by implicit up-or-out policies; (6) individuals have different abilities to perform different jobs; (7) compensation within a grade should be contingent on effort and/or performance; (8) the best matches between personnel and grades can be achieved by not motivating to move up in the ranks those who are relatively less able to perform the tasks associated with the higher grades; and (9) the pay gap across grades should be greater than the pay gap within a grade. The report points to aspects of the military that violate these guidelines and makes recommendations for future work to apply the findings.

Author   Asch, Beth J.

Title   *Navy Recruiter Productivity and the Freeman Plan*

Report#   RAND/R-3713-FMP

PubDate   1990

PubData   Santa Monica, CA: RAND

Subject   military, Navy, accession, recruiting, quality

Abstract

The Navy's pool of potential 17- to 21-year-old recruits is expected to diminish. A strategy for aiding the Navy's future recruiting effort is to alter its recruiter management techniques, particularly its incentive program, the Freeman Plan. Data from Chicago in 1986 were examined to analyze the Freeman Plan's effects on productivity. The study found that recruiting behavior is consistent with the Plan's incentives but may not be consistent with the Navy's goals. The author suggests several ways to change recruiter behavior, including increasing the point differential between high- and low-quality recruits, thereby motivating recruiters to enlist more of them; and shortening the production cycle, thereby giving recruiters less time between cycles.

Author   Asch, Beth J., and James N. Dertouzos

Title   *Educational Benefits Versus Enlistment Bonuses: A Comparison of Recruiting Options*

Report#   RAND/MR-302-OSD

PubDate   1994

PubData   Santa Monica, CA: RAND

Subject   recruiting, bonuses, accession, military, education benefits, personnel flow, compensation

Abstract

The relative cost-effectiveness of two incentive programs for recruitment—enlistment bonuses and educational benefits—are analyzed. The analysis also considers the effects of such programs on the service history of recruits, including reserve component accessions. Educational benefits are shown to significantly expand enlistment supply and increase incentives for first-term completion. Relative to bonus programs, educational benefits enhance the flow of prior service individuals into the Selected Reserve and have reduced costs because payments are deferred.

Author   Asch, Beth J., and John T. Warner

Title   *A Theory of Military Compensation and Personnel Policy*

Report#   RAND/MR-439-OSD

PubDate   1994a

PubData   Santa Monica, CA: RAND

Subject   compensation, quality, retention, pay table, intragrade pay, lateral entry, performance, promotion, personnel flow, transitioning, retirement, military, vesting

Abstract

A primary goal of military compensation is to enable the military to meet its manning objectives for force size, composition, and wartime capability. To attain these objectives, compensation must be appropriately structured to attract, retain, and motivate personnel at a reasonable cost, even when national security goals are changing. A key question facing military manpower and compensation managers is, How should military compensation be structured? Although past studies have narrowly focused on the relationship between compensation and retention, less attention has been paid to whether the military compensation system induces the best individuals to stay and seek advancements, and whether it motivates effective work. This highly technical report addresses the issue of how military compensation should be designed in light of these considerations. It presents research that helps us to develop a model of compensation in a large, hierarchical organization such as the military, a model that permits an analysis of the issues surrounding the design of military compensation. The report reaches four conclusions: (1) In a hierarchical system, pay spreads need to rise with rank to provide personnel with continuing incentives to work hard and seek promotion, and to induce the most able personnel to stay; (2) intragrade pay should be somewhat contingent upon performance and not be provided lockstep with seniority; (3) up-or-out rules are necessary to induce the separation of unpromotable personnel when pay is set administratively; and (4) retired pay may be offered for a number of reasons. The report also begins to evaluate the current military compensation system in light of the model, finding that the system appears more aimed at attracting and retaining personnel than at providing them with effective incentives to work hard and seek advancement. The need to provide incentives to perform is important in higher grades because the marginal productivity of effort rises with grade. Promotion rates decline in higher grades and are less effective as an incentive. The need for higher pay at higher grades may be offset partially by the incentives of power, privilege, and status. Increased pay differentials and the resulting increased level of competition for promotion may have negative effects in settings where high levels of teamwork and cooperation are required. The effects of limited lateral entry are examined in terms of distortions to the compensation system and the movement of personnel through the career system.

Author	Asch, Beth J., and John T. Warner
Title	*A Policy Analysis of Alternative Military Retirement Systems*
Report#	RAND/MR-465-OSD
PubDate	1994b
PubData	Santa Monica, CA: RAND
Subject	retirement, compensation, transitioning, military, cost, performance, personnel flow, quality, retention, vesting

Abstract

This report summarizes the theoretical model of the military compensation system developed in Asch and Warner (1994a), presents an empirical version of that model, and evaluates the current and alternative military retirement systems. The evaluation focuses on implications of the retirement systems for force structure, costs, and productivity. Productivity includes a system's ability to motivate individuals to work hard and effectively and to sort ability through motivating higher-ability individuals to stay and seek advancement. Using analysis of various possible revisions in retirement policy, the authors propose changes that can be expected to maintain force size and structure at no substantial change in cost. Advantages would be gains in individual effort and retention and advancement of high-ability individuals with more flexible and adaptable personnel management policies. The proposed system would vest personnel with 10 YOS in old-age annuities beginning at age 60. Pay levels and skewness of intergrade pay differentials would increase to offset the decrease in retirement benefits.

Author	Asch, Beth J., and Lynn A. Karoly
Title	*The Role of the Job Counselor in the Military Enlistment Process*
Report#	RAND/MR-315-P&R
PubDate	1993
PubData	Santa Monica, CA: RAND
Subject	recruiting, accession, military, recruit screening, job assignment, Army, personnel flow

Abstract

This report describes a theoretical framework of the enlistment process that accounts for the joint role of counselors, the person-job-match algorithm, and the supply of recruits in determining the number of enlistments, occupation, term of service and enlistment benefits chosen. Using hypotheses generated by this framework, the authors conducted empirical analysis. Results indicate that the counselor incentive plan is cost-effective in filling occupations and priority jobs when compared with educational benefits and enlistment bonuses for recruits. Previous estimates of the effect of the Army College Fund (ACF) have been overstated when the role of the job counselor has not been accounted for. Increasing counselor incentives by five points is two and one-half times more effective than educational benefits in filling an occupation and one and one-half times more effective than offering an enlistment bonus. Increasing counselor incentives is also much more cost-effective for skill-channeling. The plan would provide more effective incentives if changed to provide more challenge for the counselor and to reward counselors on the basis of individual, not group, productivity. It should be noted that the enlistment bonus and the ACF have positive market effects in addition to the skill-channeling effects examined here.

Author	Beveridge, William J.
Title	*A Network Flow and Goal Programming Approach to Modeling the Impact of Pre-Accession Training to the Trained Personnel Requirements Process*
Report#	ADA238447
PubDate	1991
PubData	Wright-Patterson Air Force Base, OH: Air Force Institute of Technology
Subject	training, accession, recruiting, civilian substitution, cost, personnel flow, Air Force, military

Abstract

HQ Air Training Command (ATC) was tasked to analyze the effect of pre-accession training to the Trained Personnel Requirement (TPR) and training production process. Pre-accession technical training during DEP allows enlistees to move directly into units after Basic Military Training. The purpose of this study was to develop a method to model the effect of pre-accession training. A network modeling and goal programming approach was used. This study has shown that a policy of pre-accession training could be cost-effective. Increases in the balk rate and subsistence package during time in training affect the savings over the current policy. Therefore, recruiting goals would have to be raised under the new policy. The number of active duty personnel retraining in skills under this policy could also reduce the savings. Earlier technical training would reduce flexibility in the system to respond to changes in demand and production in particular skill groups. The quality of contract training and effect on development of military culture were not factored into this model.

Author   Borus, Michael E., and Choongsoo Kim

Title   *Policy Findings Related to Military Service from the Youth Cohort of the National Longitudinal Surveys of Labor Market Experience*

Report#   ADA185414

PubDate   1985

PubData   Arlington, VA: Defense Manpower Data Center

Subject   military, accession, recruiting, screening, background characteristics, quality

Abstract

Some of the most policy-relevant findings from the Youth Cohort of the National Longitudinal Surveys of Labor Market Experience are summarized. Analysis of two data waves shows that minorities from better backgrounds and with better credentials are disproportionately attracted to the armed forces. Overall, the data show that the services recruit young men of high quality relative to the pool of out-of-school youth employed full time.

Author   Brown, Charles

Title   *Military Enlistments: What Can We Learn from Geographic Variation?*

Report#   ADA165663

PubDate   1984

PubData   Alexandria, VA: U.S. Army Research Institute for the Behavioral and Social Sciences

Subject   accession, recruiting, compensation, recruit screening, quality, military, Army

Abstract

This paper measures the effects of economic factors on Army enlistments of nonprior service high school graduates. Recruiting data from the Defense Manpower Data Center (DMDC) for fiscal years 1976–1982 are matched with state-level census data. The author uses a multiple-regression, pooled cross-section/time-series model to measure Army enlistments by mental category. He concludes that unemployment, pay, and the number of Army recruiters are important variables in the enlistment decision.

Author	Brown, Robert P.
Title	*The Utilization of Women by the Military*
Report#	ADA114637
PubDate	1987
PubData	Newport, RI: Naval War College
Subject	gender, personnel flow, job assignment, personnel requirements, military, development

Abstract

Current U.S. policy regarding the utilization of women in the military is inconsistent among the services. These differences are primarily the result of Title 10 of the United States Code, which prohibits women from serving on combat vessels and aircraft. Congress provided no legal barriers or restrictions regarding the placement of women on ground combat units. Demographic predictions indicate a smaller pool of qualified males available through the next two decades. Social changes and technological advancements will enhance the ability of women to expand their role in the military. Given the reluctance of Congress to return to the peacetime draft, the services will find it necessary to continue to increase the number of and expand the role of women in the military. If executed properly, it could enhance the combat readiness of the U.S. Armed Forces. A series of recommendations are contained in this document to support this view.

Author	Buddin, Richard
Title	*Enlistment Effects of the 2+2+4 Recruiting Experiment*
Report#	RAND/R-4097-A
PubDate	1991
PubData	Santa Monica, CA: RAND

Subject   recruiting, accession, educational benefits, cost, retention, personnel flow, military, Army, compensation, quality

Abstract

This report describes the enlistment effects of the Army's 2+2+4 recruiting experiment, which was aimed at attracting high-quality personnel into the active Army and encouraging their later participation in the reserves. These effects were estimated through a job-offer experiment that estimated how the program affected the recruits' choices among skills and terms of service and through a geographic experiment that assessed whether the program led to a "market expansion"—i.e., an increase in the total number of high-quality persons entering the active Army. Overall, the program seems to have accomplished its objectives for active-duty recruiting. The 2+2+4 option sold readily and benefited virtually all the occupational specialties for which it was tested. During the test, about 7 percent of all male high-quality enlistments contracts were written under the program. Moreover, the analysis indicates that the program attracted high-quality recruits into the Army and caused only a small number to change from a longer term of service to a shorter one.

Author   Buddin, Richard

Title   *Trends in Attrition of High-Quality Military Recruits*

Report#   RAND/R-3539-FMP

PubDate   1988

PubData   Santa Monica, CA: RAND

Subject   attrition, background characteristics, military, personnel flow, accession, screening, training

Abstract

This report documents attrition patterns in the U.S. military services from FY 1982 through FY 1985 and also examines attrition across training bases. The purpose of the study was to gain insight into why the recent improvements in the quality of recruits has not reduced attrition rates. Attrition patterns in recent cohorts suggest that attrition rates do not depend simply on the characteristics of individual recruits but also on other factors. These attrition patterns indicate that institutional or "demand-side" factors may play an important role in determining attrition rates. The findings indicate that the magnitude of cohort and training base effects differs by service. Service practices and policies may vary considerably at different bases and in different years. Thus, while recruit characteristics can be used to rank prospective recruits by risk category, different interpretation and enforcement of service policies seem to critically affect the actual attrition level.

Author	Buddin, Richard
Title	*Analysis of Early Military Attrition Behavior*
Report#	RAND/R-3069-MIL
PubDate	1984
PubData	Santa Monica, CA: RAND
Subject	recruiting, performance, personnel flow, accession, job assignment, costs, attrition, military, background characteristics, screening

Abstract

Analyzes the influence of preservice experiences and initial military job match on military attrition of first-term enlisted males during their first six months of service (early attrition). The dynamics of attrition behavior are examined in terms of recent firm-specific human capital and job matching models. The determinants of early attrition are compared across services and with those of civilian job separations of young workers. Some of the conclusions drawn are: enlistees with a history of frequent civilian job changes or a recent spell of unemployment are attrition-prone; aspects of the initial military occupational assignment like individual suitability and satisfaction do not significantly influence early attrition; the early attrition rate of non-high school graduates is nearly twice that of graduates even after controlling for previous work experiences, aptitude, and other variables that influence attrition; and older recruits are more attrition-prone than younger recruits.

Author   Buddin, Richard

Title   *Determinants of Post-Training Attrition in the Army and Air Force*

Report#   RAND/P-6709

PubDate   1981

PubData   Santa Monica, CA: RAND

Subject   attrition, accession, military, development, job assignment, personnel flow, background characteristics, recruit screening

Abstract

Develops a multivariate model describing the effects of individual background characteristics, duty location assignments, career turbulence, and military occupational assignments on post-training enlisted male attrition in the Army and Air Force. The research suggests that military occupation and duty location are significantly correlated with post-training attrition, after controlling for individual characteristics. The role of turbulence cannot be distinguished with current turbulence measures. Among individual characteristics, high school graduates have much lower attrition than nongraduates in all service occupational areas. Attrition does not vary significantly with mental test category, after controlling for other background and service experiences. Participation in a delayed entry program prior to entering the military reduces substantially the likelihood of post-training attrition.

Author   Buddin, Richard, and J. Michael Polich

Title   *The 2+2+4 Recruiting Experiment: Design and Initial Results*

Report#   ADA239671, RAND/N-3187-A

PubDate   1990

PubData   Santa Monica, CA: RAND

Subject   recruiting, accessions, personnel flow, term length, quality, skill needs, educational benefits, Army, military

Abstract

This Note describes the design and first six months of experience for a national experiment on a proposed new recruiting program for the U.S. Army. The program, called the "2+2+4" recruiting option, is one of the tools the Army believes could help sustain its ability to attract high-quality young people during difficult recruiting periods in the future. The authors present RAND's design for the test as a controlled experiment, similar to earlier enlistment incentive tests, and present preliminary tabulations of results during the first six months of the test. The test established a framework for systematic assessment of the 2+2+4 program and set up a precise mechanism for possible future tests of other enlistment options through individually randomized assignment in the REQUEST system. The test showed that a substantial number of recruits are willing to commit for two years in the Selected Reserve to obtain an Army College Fund benefit. It also showed that offering the 2+2+4 option has led relatively few recruits to choose a short term of service in place of a longer term or to move from a combat to a non-combat skill. It is too soon to determine whether the program led to a significant increase in the total number of high-quality recruits entering the Army.

Author	Buddin, Richard, Daniel S. Levy, Janet M. Hanley, and Donald M. Waldman
Title	*Promotion Tempo and Enlisted Retention*
Report#	RAND/R-4135-FMP
PubDate	1992
PubData	Santa Monica, CA: RAND
Subject	military, promotion, retention, personnel flow, quality, performance, development, Army, Air Force

Abstract

Previous retention research has concentrated on military/civilian pay levels and has largely ignored changes in military promotion timing. Over the past several years, promotion tempo has slowed considerably in the enlisted force; the implications of the slowdown, however, have received little attention. This report examines factors that affect promotion timing during the first enlistment term and ex-

amines how changes in promotion tempo affect the first-term retention decision. The authors developed and estimated a joint, integrated model of promotion and first-term retention behavior and
compared the results with those from previous approaches that use
little (if any) promotion information. The results demonstrate that
retention models are sensitive to the specification of individual promotion opportunities at the end of the first term. The approach also
shows that several key parameters of traditional models have been
misleading because they have not adjusted for promotion timing.
The authors conclude that soldiers are quite sensitive to promotion
tempo and that promotion could be used to complement military
pay and bonus policies in retaining quality personnel in hard-to-fill
skills. Promotion policy should be an important part of any compensation package. Finally, they suggest that the services not rely on reduced promotion tempo to induce lower retention during the
planned military drawdown.

Author   Callander, Bruce D.

Title   "A New Shot at the Promotion System"

Report#

PubDate   1995

PubData   *Military Life*, July 5, 1995, pp. 70–73.

Subject   promotion, personnel flow, military, Air Force

Abstract

This article reviews proposed changes to officer and enlisted promotion systems in the Air Force and discusses perceptions of the systems revealed in opinion surveys. Officers rate promotion opportunity as the most dissatisfying feature of Air Force life, while enlisted
members rank it as the third biggest dissatisfier. Most criticism centers on the evaluation systems in use.

Author   Campbell, John P., and Lola M. Zook

Title   *Building and Retaining the Career Force: New Procedures
              for Accessing and Assigning Army Enlisted Personnel:
              Final Report*

Report#   ADA320954

PubDate   1996

PubData   Alexandria, VA: Human Resources Research
             Organization

Subject   accession, military, performance, screening, job match,
             retention

Abstract

The Career Force research project is the second phase of a two-phase
U.S. Army program to develop a selection and classification system
based on expected future performance for enlisted personnel. In the
first phase, Project A, a large and versatile data base was collected
from a representative sample of military occupational specialties and
used to (a) validate the Armed Services Vocational Aptitude Battery
and (b) develop and validate new predictor and criteria measures
representing the entire domain of potential measures. Building on
this foundation, Career Force research will finish developing the se-
lection/classification system and evaluate its effectiveness, with emphasis on assessing second-tour performance.

Author   Carpenter-Huffman, Polly

Title   *The Cost-Effectiveness of On-the-Job Training*

Report#   RAND/P-6451

PubDate   1980

PubData   Santa Monica, CA: RAND

Subject   training, cost, Air Force, military, performance,
            development

Abstract

A discussion of ways to assess the cost and effectiveness of on-the-
job training (OJT), which is currently (1980) undergoing intensive
study by the Air Force. The paper highlights some of the conceptual
and analytical difficulties in making such assessments, pointing out
some problems that do not arise in analyzing other types of training
processes. Although OJT presents special challenges for cost-effec-

tiveness analysis, these challenges are well worth meeting in view of the growing scarcity of skilled people in the military services.

Author	Carter, Grace M., and Rachelle Kisst Hackett
Title of	"The Effect of Selective Reenlistment Bonuses on Terms Enlistment in the Air Force"
Report#	RAND
PubDate	unpublished
PubData	Santa Monica, CA: RAND
Subject	retention, bonuses, compensation, personnel flow, Air Force, military, development

Abstract

This unpublished draft addresses the question, "To what extent do bonuses encourage longer Air Force reenlistment contracts and hence greater expected manyears of service?" It describes regression models that relate the number of years in the contract to character-istics of the airman and his service environment—particularly the Selective Reenlistment Bonus. Separate models were fit for persons eligible for zone A, zone B, and zone C bonuses. Results from both linear and logistic regression models are presented and discussed. The amount of the Selective Reenlistment Bonus offered in the reen-listee's specialty is one of the strongest determinants of term of en-listment (TOE). Other determinants of TOE include the length of time the reenlistee has already served in the Air Force, occupation, and demographic characteristics.

Author	Celeste, Jeanna F.
Title	*Delayed Entry Program Contracting Cohort Loss Analysis: A Replication*
Report#	ADA182408
PubDate	1985
PubData	Alexandria, VA: U.S. Army Research Institute for the Behavioral and Social Sciences

Subject   recruiting, accessions, personnel flow, DEP, attrition, military, Army

Abstract

This paper presents the approach and findings of a Delayed Entry Program contract loss analysis using a cohort methodology. Contracts written from October 1980 to March 1983 were studied. This analysis was conducted as a replication of work performed earlier by the U.S. Army Recruiting Command's DEP Efficiency Task Force. Individual contract-level data on characteristics such as time in the DEP, AFQT groups, educational level, and gender were examined for their relationship to contracting cohort loss rates. Length of time in DEP was positively related to the loss rate. Male high school graduates had lower loss rates. Female contractees experienced loss rates more than twice the rates of males.

Author   Congressional Budget Office

Title   *Quality Soldiers: Costs of Manning the Active Army*

Report#

PubDate   1986

PubData   Washington, DC: Congressional Budget Office

Subject   military, costs, Army, accessing, quality, screening

Abstract

Every year the Army recruits about 125,000 young men and women into its active duty force. Despite some trends that may make recruiting more difficult—higher rates of employment and declining numbers of military age youths—the Army should continue to attract well-qualified candidates. Under CBO's baseline projection, well over 80 percent of the Army's male recruits over the next five years will hold high school diplomas. Army proposals to raise this rate to 90 percent are examined. The added costs of recruitment more than offset any savings realized due to lower dropout rates and reductions in training costs.

Author   Cooke, Timothy W.

Title   *Evaluation of the Targeted Enlistment Bonus (TEB) for Nuclear Field Recruits*

Report#   ADA191786

PubDate   1987

PubData   Alexandria, VA: Center for Naval Analyses

Subject   accession, bonuses, recruiting, personnel flow, cost, Navy, military, compensation

Abstract

This research memorandum contains the last of the three evaluations of the Targeted Enlistment Bonus for Nuclear Field recruits. The TEB differs from previous enlistment bonuses by varying the bonus amounts according to the season a recruit begins active duty. Historically, Nuclear Field accessions have been characterized by a seasonal surge in the summer months, reflecting the presence of many Nuclear Field recruits for beginning service shortly after obtaining a high school diploma. The TEB is designed to assist recruiters in achieving a more level flow of accessions during the year. It was tested during the 18-month period from October 1985 through March 1987. For the evaluation, Nuclear Field recruits during this period are compared to those of previous years in terms of the phasing of accessions and enlistment contracts, and indicators of recruit quality. Savings associated with the TEB experiment are calculated, and implications for potential changes in the TEB are drawn. The TEB was successful in smoothing seasonal changes in accessions. Use of TEB would allow the size of enlistment bonuses to be reduced with no change in recruit quality.

Author   Cooper, Deanna L.

Title   *A Cross-Sectional Investigation of the Effects of Regional Labor Market Conditions on the Reenlistment Decisions of Air Force Enlistees*

Report#   ADA246640

PubDate   1991

PubData   Wright-Patterson Air Force Base, OH: Air Force Institute
of Technology

Subject   military, retention, Air Force

Abstract

Using survey data collected in 1984 and reenlistment data collected
in 1986, a cross-sectional investigation of the effects of general labor
market conditions, as measured by regional unemployment rates, on
the reenlistment decisions of first-term Air Force enlistees was conducted. Additionally, the effects of labor market perceptions, cognitive ability, and tenure on reenlistment decisions were studied. A
model of voluntary employee turnover was developed and tested.
One of seven hypotheses was supported, providing little support for
the proposed model. Unemployment, labor market perceptions, and
cognitive ability were not found to affect reenlistment decisions.

Author   Cymrot, Donald J.

Title   *Early Attrition in FY 1985: The Effects of the Delayed*
Entry   *Program, Accession Month, and Enlistment Program*

Report#   ADA180137

PubDate   1986

PubData   Alexandria, VA: Center for Naval Analyses

Subject   attrition, military, accession, DEP

Abstract

This research memorandum examines the effects of three factors on
attrition from the Navy within 2 and 6 months of shipping for recruits who entered in FY 1985. These three factors are participation
in the Delayed Entry Program, month of shipment, and enlistment
program. The results indicate that DEP recruits have lower attrition
than direct shippers, that attrition among direct shippers is higher in
months with high accession rates, and that attrition rates vary by enlistment program.

Author   Daula, Thomas V., and Robert A. Moffitt

Title   "Estimating a Dynamic Programming Model of Army
              Reenlistment Behavior"

Report#

PubDate   1991

PubData   Alexandria, VA: U.S. Army Research Institute for the
              Behavioral and Social Sciences (in *Military
              Compensation and Personnel Retention,* edited by C. L.
              Gilroy, D. K. Horne, D. A. Smith)

Subject   military, Army, reenlistment, accessing, screening,
              retention

Abstract

This paper demonstrates a method for estimating a simple dynamic
programming model. In a simple model of optimal stopping behavior, the authors demonstrate an iterative estimation method in which
the solution to the dynamic program alternates with a simple probit
estimation of a stopping function. The solution to the dynamic program thus takes place outside the probit estimation step, simplifying
the estimation procedure and providing savings in computational
effort.

Author   Department of Defense

Title   *8th Quadrennial Review of Military Compensation*

Report#

PubDate   1997

PubData   Washington, DC: Department of Defense

Subject   military, compensation, costs, retirement, planning,
              management, performance

Abstract

Leaders manage their human resources to obtain organizational
performance by channeling individual potential into organizational
achievement. Research cited throughout this report demonstrates
that fundamental—strategic—choices about organizing, managing,

and rewarding people affect organizational performance. The 8th Quadrennial Review of Military Compensation differs from previous reviews in that it was tasked to take a strategic approach. Rather than focusing on specific issues or perceived problems with the existing compensation system, this review sought to develop a process to align all the elements of the human resource management system to obtain the maximum effect from the resources devoted to people.

Author   Department of Defense

Title   *Forum on Strategic Human Resource Management*

Report#

PubDate   1996

PubData   Washington, DC: Department of Defense

Subject   strategic management, compensation, planning, military

Abstract

This report provides a synopsis of a February 14, 1996, forum hosted by Secretary of Defense William Perry, which sought to determine how strategic human resource management can contribute to achieving the services' roles and missions into the 21st century. Attendees included senior leaders from DoD and the uniformed services and chief executive officers of five major American corporations. Comments and suggestions are summarized in five categories: (1) strategic management, (2) strategic human resource management, (3) a strategic compensation perspective, (4) the changing environment and future requirements, and (5) purposeful, strategic change.

Author   Department of Defense

Title   *Managing Enlisted Seniority*

Report#

PubDate   1988

PubData   Washington, DC: Department of Defense

Subject   military, transitioning, personnel flow, grade content,
                 development, promotion

Abstract

There are indicators that emerging seniority patterns might be inappropriate. While seniority growth is frequently viewed as an opportunity to increase readiness, there are no consistent nor explicit measures employed within the department in gauging the cost-effectiveness of such seniority. This study's central recommendation is the
identification of cost-effective seniority benchmarks against which
the grade patterns (derived form the manpower requirements process) would be evaluated.

Author   Department of Defense

Title   *Enlisted Personnel Management System*

Report#   ADA269471

PubDate   1984

PubData   Washington, DC: Department of Defense

Subject   personnel system, military

Abstract

This Directive outlines the goals, objectives, and responsibilities of
the enlisted personnel management system.

Author   Dertouzos, James N.

Title   *Recruiter Incentives and Enlistment Supply*

Report#   RAND/R-3065-MIL

PubDate   1985

PubData   Santa Monica, CA: RAND

Subject   accession, recruiting, Army, military, personnel flow,
                recruiting characteristics, high-quality personnel

Abstract

In an empirical study of Army recruiting data, RAND concluded that demand factors such as recruiter quotas and incentives to achieve and exceed them play a critical role in the determination of enlistments. Recruiters who achieve high-quality quotas are less likely to be induced by existing incentives to increase their productivity than are those who do not achieve high-quality quotas. Thus, resource expenditures meant to induce an increase in potential supply may not result in actual high-quality enlistments because recruiters do not have incentives to secure them. Two major research and policy implications emerge: (1) Future attempts to project enlistments or to analyze the role of supply factors must consider demand factors explicitly; (2) the effectiveness of resource expenditures can be enhanced dramatically if appropriate incentives exist for recruiters.

Author	Dertouzos, James N.
Title	*Enlistment Supply, Recruiter Objectives, and the All-Volunteer Army*
Report#	RAND/P-7022
PubDate	1984
PubData	Santa Monica, CA: RAND
Subject	recruiting, accession, Army, military, quality, cost, personnel flow

Abstract

This paper explicitly considers the interaction of demand factors and variables characterizing supply in the determination of enlistment outcomes. See Dertouzos (1985), RAND/R-3065-MIL, for additional results of this research.

Author	Doering, Zahava D., and David W. Grissmer
Title	*Active and Reserve Force Attrition and Retention: A Selected Review of Research and Methods*
Report#	RAND/P-7007
PubDate	1985

PubData   Santa Monica, CA: RAND

Subject   accession, attrition, military, personnel flow, retention, compensation

Abstract

This paper reviews research on the dual issues of attrition (separation prior to the completion of agreed-upon terms of military service) and retention (individuals' voluntary decisions to remain in the military for additional terms of service) for the U.S. active and reserve forces. The study is limited to the enlisted force only. It describes the changing composition of U.S. active and reserve personnel, summarizes research and research methods used to study attrition and retention for both the active and reserve components, discusses different methodological issues, and notes the implications for future research in both the United States and NATO countries.

Author   Elig, Timothy W., R. M. Johnson, P. A. Gade, and Allyn Hertzbach

Title   *The Army Enlistment Decision: An Overview of the ARI Recruit Surveys, 1982 and 1983*

Report#   ADA164230

PubDate   1984

PubData   Alexandria, VA: U.S. Army Research Institute for the Behavioral and Social Sciences

Subject   military, Army, accession, recruiting, background characteristics

Abstract

This paper describes a program of research on enlistment decision-making. Initial efforts—surveys of recruits about the factors that encourage, discourage, and drive their enlistment decision—are described in detail. Two surveys of over 25,000 recruits have been conducted by the Army's Manpower and Personnel Research Laboratory. Comparisons are made between the 1979, 1982, and 1983 surveys. Examinations of incentives show that net Army gains from every 100 two-year enlistments is estimated to be 71 years of service. Net gains from every 100 ACF enlistments is estimated to be enlistments in hard-to-fill MOSs of 35 high-quality males who would otherwise not have enlisted.

Author   Everage, Henry B.

Title   *Paygrade Structure and Flows into the "Compressed" Ratings*

Report#   ADA217665

PubDate   1989

PubData   Alexandria, VA: Center for Naval Analyses

Subject   military, development, promotion, compensation

Abstract

This research memorandum examines the billet authorizations on the Enlisted Billet File to gain some insight into paygrade distributions and how these distributions relate to promotion opportunity. Of particular interest were the compressed ratings that occur at the senior paygrades. Comparisons of promotion opportunities were made among ratings and between compressed and noncompressed rating groups.

Author   Farris, H., W. L. Spencer, J. D. Winkler, and J. P. Kahan

Title   *Computer-Based Training of Cannon Fire Direction Specialists*

Report#   RAND/MR-120-A

PubDate   1993

PubData   Santa Monica, CA: RAND

Subject   training, cost, Army, military, development

Abstract

Using the advanced individual training of Cannon Fire Direction Specialists as a case study, this report identifies alternative approaches for individual training and analyzes their cost implications. The study suggests that the current course can be reorganized to reduce course length and conserve resources while meeting fundamental training objectives. Specifically, 20 percent of the current training time contains tasks that may not be performed in the subsequent duty assignment. The analysis further identifies tasks well-suited for computer-based training (CBT). These tasks, which cover fire detection center and fire mission operations, require complex computational and diagnostic skills that are hard to train and, thus, lend themselves to individualized CBT instruction. If CBT were implemented along with other steps to realign the course, additional savings in training manpower and costs could be realized. Although the cost of courseware development will affect the savings, a payback period of three years should prove economically justified given the continuing battlefield requirement for technical support to fire missions.

Author   Fernandez, Judith C., and Dennis De Tray

Title   *Sources and Characteristics of Prior Service Accessions:*
              Evidence from One Cohort

Report#   RAND/N-2064-MIL

PubDate   1984

PubData   Santa Monica, CA: RAND

Subject   recruiting, accession, personnel flow, cost, military,
              development

Abstract

Enlisted military personnel who leave the active forces form a valuable pool of trained personnel from which come entrants to the Selected Reserve and the Individual Ready Reserve. In addition, members of this pool may later return to active duty, where they may function as alternatives to nonprior service accessions or to reenlistments. To better understand this source of trained manpower, the authors followed the 1974 cohort of active duty enlistees over time, and investigated the flows among three components of the U.S. armed forces—active duty, Selected Reserve (SR), and Individual Ready Reserve (IRR)—and the flows between the civilian and military sectors. The findings indicate there is a large untapped pool of potential prior service accessions from which to attract personnel into the active forces, and little evidence that drawing from this pool will divert trained personnel from the Selected Reserve.

Author	Fernandez, Judith, C., and J. H. Kawata
Title	*A System for Allocating Selective Reenlistment Bonuses*
Report#	RAND/N-2829-FMP
PubDate	1989
PubData	Santa Monica, CA: RAND
Subject	compensation, bonuses, military, costs, retention, personnel flow, development

Abstract

This Note explores the conceptual basis for a cost-minimizing approach for allocating Selective Reenlistment Bonuses among military occupational specialties that are of varying degrees of criticality to the defense mission. It also presents an operational methodology for implementing that approach in the face of the data and computational limitations that characterize the environment in which bonuses are allocated. The concept and methodology support a personal-computer-based program that yields a recommended allocation of bonuses. Appendixes describe the details of the computer software.

Author	Fernandez, Richard L., and Jeffrey B. Garfinkle
Title	*Setting Enlistment Standards and Matching Recruits to Jobs Using Job Performance Criteria*
Report#	RAND/R-3067-MIL
PubDate	1985
PubData	Santa Monica, CA: RAND
Subject	recruiting, performance, personnel flow, accession, job assignment, costs, attrition, military

Abstract

Every year the military services are faced with the tremendous tasks of choosing 300,000 new recruits from among the larger number who are willing to serve, and of deciding in which specialty each of the 300,000 should be trained. This report describes a study largely concerned with determining whether there is any objective basis for enlistment standards and for matching recruits to jobs. It develops a cost/performance tradeoff model that appears to be a useful tool for setting job standards and for prescribing appropriate service-wide standards, but finds that three questions must be answered before the model can be used objectively. A key element of the model is the "qualified man-month," a single performance measure that combines attrition and job performance information. A related RAND report by D. J. Armor, R. L. Fernandez, et al. (*Recruit Aptitudes and Army Job Performance: Setting Enlistment Standards for Infantrymen*, R-2874) provides useful background to this report.

Author	Firestone, J. M., and R. J. Harris
Title	"Sexual Harassment in the U. S. Military: Individualized and Environmental Contexts"
Report#	
PubDate	1994
PubData	*Armed Forces and Society*, Vol. 21, No. 1, pp. 25–43.
Subject	military, gender, development, harassment

Abstract

This article presents the results of an analysis of the "1988 DoD Survey of Sex Roles in the Active-Duty Military." A sample of 20,249 was taken from the Army, Navy, Marines, Air Force, and Coast Guard. Two forms of sexual harassment were examined. Environmental harassment relates to the existence of a hostile work environment. Individual harassment is more personal and includes quid pro quo demands or unwanted advances on particular individuals. Results show a pervasive pattern of sexual harassment, usually of women, that spans rank and work site contexts. Environmental harassment context is highly predictive of individualistic harassment experiences. In settings where no environmental harassment was reported, 99 percent of respondents had not experienced individual experiences of harassment. In settings where environmental harassment was reported over 75 percent had experienced individual harassment.

Author	General Accounting Office
Title	*Military Attrition: DOD Could Save Millions by Better Screening Enlisted Personnel*
Report#	GAO/NSIAD-97-39
PubDate	1997
PubData	Washington, DC: U.S. General Accounting Office
Subject	military, accession, recruiting, screening, cost

Abstract

Pursuant to a congressional request, GAO reviewed the attrition rates of first term, active duty military personnel who are separated within the first six months of their enlistments, focusing on (1) how much the services could save by achieving their goals for reducing six-month attrition; (2) the adequacy of the data that the of the DoD uses to allow it to establish realistic goals for reducing attrition; and (3) the principal reasons that enlistees are separated from the services while they are still in training. GAO found that (1) all the services agree that reducing attrition is desirable; (2) three services have attrition-reducing targets ranging from 4 to 10 percent; (3) if the services reach their goals, they would realize immediate short-term annual savings ranging from around $5 million to $12 million; (4) the services may not be able to realize savings through reductions in their related training and recruiting infrastructure for many years, but possible long-term savings could range from more than $15 million to $39 million; (5) despite the fact that the services have these goals, DoD, at present, lacks consistent and complete information on the causes of attrition; (6) implementing arbitrary attrition-reduction goals could result in a reduction in the quality of recruits; (7) DoD's primary database for managing attrition cannot be used to adequately determine the reasons that recruits separate and to set appropriate targets for reducing attrition for two reasons: (a) The services interpret and apply DoD's uniform set of separation codes differently because DoD has not issued directives on how to interpret them; and (b) current separation codes capture only the official reason that an enlistee leaves the service; (8) thousands of recruits are separated in the first six months because the services do not adequately screen applicants for disqualifying medical conditions or for preservice drug use; (9) one reason that this screening is inadequate is that recruiters do not have sufficient incentives to ensure that their recruits are qualified; (10) thousands of recruits also are separated who fail to meet minimum performance criteria; and (11) recruits have problems meeting performance standards because they are not physically prepared for basic training and because they lack motivation.

Author   General Accounting Office

Title   *Military Retirement: Possible Changes Merit Further Evaluation*

Report#   GAO/NSIAD-97-17

PubDate   1997

PubData   Washington, DC: U.S. General Accounting Office

Subject   military, retirement, transitioning, cost, personnel flow, compensation, vesting

Abstract

Pursuant to a congressional request, GAO reviewed selected aspects of the military retirement system, focusing on (1) military retirement costs; (2) the role of military retirement in shaping and managing U.S. forces; and (3) proposed changes to modernize the system and contribute to more efficient force management. GAO found that (1) payments from the military retirement fund to military retirees and their survivors have been rising over several decades as both the number of military retirees and the average payment to individual retirees have increased; (2) these payments are expected to peak at slightly more than $30 billion in 2007; (3) DoD annual budgetary costs have declined because of lower benefits for new entrants, changes in economic and actuarial assumptions to reflect experience, and recent decreases in force size; (4) the military retirement system provides an increasing incentive for service members to stay in the military as they approach 20 years of service and encourages them to leave thereafter, helping DoD to retain mid-career personnel and yielding a relatively young force; (5) the system can also impede effective force management because military personnel with fewer than 20 years of service are not entitled to any retirement benefits; (6) the services have been reluctant to involuntarily separate personnel with fewer than 20 years of service, beyond a certain point, due to financial consequences for service members and their families and the resulting effect on morale; (7) some analysts believe the military retirement system is an obstacle to achieving a force of the right size and composition because the system provides the same career length incentive for all categories of personnel; (8) proposals to change the military retirement system range from modifications of various features of the current system to more fundamental changes to the retirement system; and (9) earlier vesting of at least a portion of military retirement benefits is a common feature of proposed changes.

Author	General Accounting Office
Title	*Physically Demanding Jobs: Services Have Little Data on Ability of Personnel to Perform*
Report#	GAO/NSIAD-96-169
PubDate	1996a
PubData	Washington, DC: U.S. General Accounting Office
Subject	military, gender, recruit screening, accession

Abstract

Pursuant to a congressional request, GAO reviewed the use and development of gender-neutral occupational performance standards in the military services, focusing on how the services implement and evaluate standards. GAO found that (1) each service takes a different approach to screening members' physical fitness; (2) the Air Force is the only service that requires new recruits to take a strength aptitude test; (3) the Air Force uses the results to qualify individuals for their military occupations; (4) the services believe that their approaches to assigning members to physically demanding tasks are appropriate, because they receive few complaints from members about such tasks; (5) the services have few data to assess a members' capability to perform tasks; (6) the Army has systematically collected physical performance data since 1989; (7) the data show that at least 84 percent of the Army members had no problems in completing their tasks; (8) a 1994–1995 survey determined that 51 to 79 percent of members have no problem in completing physically demanding tasks; and (9) the validity of the Air Force's strength aptitude test is questionable because of concerns about the administration, accuracy, and relevance of the tests' physical requirements.

Author   General Accounting Office

Title   *Military Personnel Reassignments: Services Are Exploring Opportunities to Reduce Relocation Costs*

Report#   GAO/NSIAD-96-84

PubDate   1996b

PubData   Washington, DC: U.S. General Accounting Office

Subject   military, development, cost, duty assignment

Abstract

In fiscal year 1995, the military spent nearly $3 billion to move 850,000 service members and their families. GAO has found that few opportunities exist to reduce the costs of permanent change-of-station moves. Overseas commitments and other laws also require the military to move many service members each year. Despite these constraints, the military is trying to cut annual costs by reducing the number of permanent change-of-station moves. To further reduce costs, the services are encouraging consecutive assignments in some geographic areas and increasing tour lengths where possible. Finally, the Defense Department can further decrease its overseas military requirements by hiring overseas contractors. The number of relocations, but not their costs, decreased in proportion to the defense downsizing from fiscal year 1987 through fiscal year 1995. The main reasons that permanent change-of-station costs did not decrease were inflation, changes in some entitlements, and an increase in the number of service members with dependents. According to military officials, the frequency of permanent change-of-station moves is only a minor contributor to readiness problems in military units. Other factors, especially the increase in deployments for operations other than war, have a greater effect on readiness.

Author	General Accounting Office
Title	*Trends in Active Military Personnel Compensation Accounts for 1990–97*
Report#	GAO/NSIAD-96-183
PubDate	1996c
PubData	Washington, DC: U.S. General Accounting Office
Subject	compensation, military, cost

Abstract

This report reviews trends in active military personnel compensation from 1990 to 1997. During this period the total military personnel budget for active forces is expected to fall by 30 percent, which closely follows a similar reduction in total personnel levels.

Author	General Accounting Office
Title	*Basic Training: Services Using a Variety of Approaches to Gender Integration*
Report#	GAO/NSIAD-96-153
PubDate	1996d
PubData	Washington, DC: U.S. General Accounting Office
Subject	military, accession, gender, cost, quality, basic training

Abstract

Pursuant to a congressional request, GAO reviewed the services' enlisted basic training programs to determine (1) the extent to which the services conduct gender-integrated basic training; and (2) changes that the services made to accommodate gender-integrated training. GAO found that (1) the military services use several different approaches to integrate men and women into basic training programs; (2) male and female trainees experience the same type of recruit program outside of having separate berths, medical examinations, hygiene classes, and physical standards; (3) the costs associated with gender-integrated training are low; (4) the Army spent $67,000 to modify its barracks to house gender-integrated basic training units; (5) these modifications included installing partitions between male and female berthing areas and creating separate bathrooms for male and female trainees; (6) the Navy made modifications to its basic training facility in response to base realignment and closure decisions; (7) the performance of military trainees is not harmed by gender-integrated training programs; (8) women performed better in the Army's gender-integrated training units, but male trainees' performance remained the same; and (9) the Army has no record of the gender-integrated training it conducted during the late 1970s and early 1980s to compare with its current program.

Author   General Accounting Office

Title   *Retention Bonuses: More Direction and Oversight Needed*

Report#   GAO/NSIAD-96-42

PubDate   1995a

PubData   Washington, DC: U.S. General Accounting Office

Subject   military, development, bonuses, retention, transitioning, cost, skill needs, compensation

Abstract

The Selective Reenlistment Bonus Program was created 30 years ago to help the military retain highly skilled service members. However, GAO found that in 1994, the DoD paid about $64 million in retention bonuses to individuals who worked in job categories in which positions were filled or who had been paid incentives to leave. Military officials defended their management of the retention and separation incentive programs, asserting that each is targeted at different segments of the force, that retention and separation incentives went to personnel in different grades and year groups, and that payment of separation incentives did not mean that they were satisfied with manning levels. GAO believes that if a skill is experiencing shortages that warrant the payment of retention incentives, it is not prudent to pay incentives to others with those same skills to leave the service.

The Office of the Secretary of Defense is not providing adequate direction and oversight of the programs. GAO reviewed whether the DoD has effectively managed the Selective Reenlistment Bonus (SRB) Program. GAO found that (1) the services have awarded some SRBs to personnel in high-skill categories where a high percentage of the required positions are already filled; (2) in FY 1994, 43 percent of the new SRB contracts went to service members in skill categories where 90 percent or more of the required positions were filled and in which many higher skill level service members were paid incentives to leave the service; (3) each SRB program is targeted to different segments of the military, including personnel in different grades and year groups; and (4) the Office of the Secretary of Defense has not provided adequate oversight of the SRB program, having performed only one skills review in FY 1991.

Author   General Accounting Office

Title   *Equal Opportunity—DoD Studies on Discrimination in the Military*

Report#   GAO/NSIAD-95-103

PubDate   1995b

PubData   Washington, DC: U.S. General Accounting Office

Subject   military, gender, equal opportunity

Abstract

We identified 72 studies, dating from 1974 to 1994, related to the issue of equal opportunity in the military. We categorized the studies, based on their content, into the areas of equal opportunity climate, training, sexual discrimination and harassment, promotions, discipline, and recruitment. The following are some of the general observations made in the studies:

1. Blacks and women tended to hold negative perceptions regarding equal opportunity in the military. Poor training and lack of visible chain of command participation led to decreased emphasis on the Army's equal opportunity program.

2. Racial harmony training in the Army improved effectiveness in dealing with racial problems.

3. Human relations training in the Air Force seemed to give sufficient attention to service-specific issues and applications.

4. Sexual harassment is a problem in all services, and efforts to prevent it have not been totally effective. Most victims did not take formal action because they anticipated a negative outcome.

5. Performance ratings and fitness reports of women serving in the Navy contained gender-type language that may have negatively affected their career paths and opportunities for promotion.

6. When compared to their white counterparts, black service members were overrepresented in courts-martial with respect to certain types of offenses.

7. White males are likely to continue to make up the majority of service members. Hispanic males will probably not increase their representation in the military despite the relatively rapid growth of the Hispanic population because their service eligibility rates are lower than those of white males.

8. The military services reported taking complete or partial action on 26 of the 38 studies that contained recommendations. They could not provide information on the status of the recommendations in the other 11 reports. The recommendation in one report is obsolete due to a change in policy. In addition, the cognizant organizations could not locate or provide copies of three reports.

Author   General Accounting Office

Title   *Military Equal Opportunity: Certain Trends in Racial and Gender Data May Warrant Further Analysis*

Report#   GAO/NSIAD-96-17

PubDate   1995c

PubData   Washington, DC: U.S. General Accounting Office

Subject   gender, equal opportunity, military, accession

Abstract

This report reviews Military Equal Opportunity Assessments (MEOA) prepared by the military services and the use of those assessments by the Office of the Secretary of Defense. Certain active duty personnel data were examined to determine whether possible racial or gender disparities in selection rates exist. Gender and racial disparities were found, but limitations of the data collection methods across the services and the scope of the review prevented identification of the causes of disparities. More analysis is needed before making conclusions about DoD's personnel management practices.

Author	General Accounting Office
Title	*Military Recruiting: More Innovative Approaches Needed*
Report#	GAO/NSIAD-95-22
PubDate	1994
PubData	Washington, DC: U.S. General Accounting Office
Subject	military, recruiting, accession, cost, attrition

Abstract

The military has overstated its future recruiting challenges—the number of eligible high school graduates is actually expected to grow steadily at least through the year 2000—and current recruiting operations are plagued by inefficiencies, with half the network of 6,000 recruiting offices supplying just 13.5 percent of enlistees. Although many studies have suggested ways to consolidate or eliminate layers of management to cut costs, the services have been reluctant to change existing organizational structures. Moreover, the military services plan to add more recruiters even though fewer recruits are needed to support the smaller military force resulting from downsizing. The declining propensity of youth to enlist may not be a good indicator of supply levels since more than half of all recruits come from groups with negative intentions to join the military. DoD and the services have not fully implemented proven programs that can reduce first-term attrition and decrease the number of recruits needed each year.

Author	General Accounting Office
Title	*Military Downsizing: Balancing Accessions and Losses Is Key to Shaping the Future Force*
Report#	GAO/NSIAD-93-241

PubDate   1993

PubData   Washington, DC: U.S. General Accounting Office

Subject   military, downsizing, accession, transitioning, readiness

Abstract

Although the military services have significantly reduced accession levels over those of previous years, they are also recruiting large numbers of personnel to better ensure a balanced force across the various pay grades and skill areas. Congress has prescribed reduction targets and provided other guidance to the Pentagon to facilitate downsizing, minimize involuntary separations, and preserve a balanced force. This report examines the DoD's adherence to congressional guidance and authorization in military downsizing. GAO discusses (1) what progress DoD has made toward meeting reduction targets, (2) how downsizing actions are affecting new recruiting or accessions, (3) what range of voluntary and involuntary reduction actions are being taken to meet downsizing objectives, (4) how downsizing is being accomplished across various groupings of officer and enlisted personnel by years of service and how this is affecting future force profiles, and (5) what issues might be important to future reduction decisions.

Author   General Accounting Office

Title   *Enlisted Force Management: Past Practices and Future Management*

Report#   GAO/NSIAD-91-48

PubDate   1991

PubData   Washington, DC: U.S. GPO

Subject   force management, Air Force, Army, personnel flow, military, cost, development, promotion

Abstract

This report discusses enlisted force management in the Department of Defense. The review was conducted because the high cost of maintaining a balanced and ready enlisted force underscores the need for efficient management of these resources, particularly as the services transition to smaller forces. Specifically, the report covers how the Army and Air Force (1) manage the size and composition of their enlisted forces, (2) plan for enlisted force reductions, and (3) comply with enlisted force management requirements. While the Army and Air Force have complied with DoD enlisted force management requirements in most areas, both have exceeded limits on career content (enlisted career personnel with more than four years of service). Enlisted seniority has been on an upward trend throughout the 1980s. This has resulted in increased personnel costs in terms of military pay and retirement benefits. Not managing career content within targets also creates cycles of peaks and valleys as the services are forced to bring in fewer recruits to stay within end-strength limits.

Author	Gilbert, Ronald G.
Title	"Human Resource Management Practices to Improve Quality: A Case Example of Human Resource Management Intervention in Government"
Report#	
PubDate	1991
PubData	*Human Resource Management,* Vol. 30, No. 2, pp. 183–198
Subject	civilian, Total Quality Management (TQM), work systems, employee involvement, best practices, cost, compensation, performance, productivity, job satisfaction, incentives, flexibility, development

Abstract

This article describes the quality approach to management in government and contrasts it with the classic, often more bureaucratic, approach. It places particular focus on Project Pacer Share, a government total quality management effort under way at McClellan Air Force Base where, using quality tools and techniques, major reforms in the U.S. Civil Service System and human resource management are being tested. Specific human resource management initiatives that need to be undertaken to support organization-wide quality performance are presented. Innovative practices included gainsharing, job consolidation, pay banding, cross-training, flexible assignments, and elimination of individual performance appraisal in favor of team appraisals. Preliminary results show that the innovative practices associated with the quality management approach had positive effects on customer service, labor-management relations, work processes, job satisfaction, and teamwork. A 30 percent decrease in the number of civilians and supervisors was realized along with a payroll savings of $7 million.

Author Gilroy, Curtis L., David K. Horne, and Alton Smith, editors

Title *Military Compensation and Personnel Retention: Models and Evidence*

Report#

PubDate 1991

PubData Alexandria, VA: U.S. Army Research Institute for the Behavioral and Social Sciences

Subject retention, compensation, quality, bonuses, SRB, retirement benefits, educational benefits, attrition, military, Army, accession

The research presented in this volume is at the cutting edge of the theoretical and empirical literature on labor supply, and provides the U.S. Army with the basis upon which to make significant decisions to improve the management of enlisted retention. The role of compensation and other factors in the individual reenlistment decision has been the topic of considerable interest and research. In this volume we have brought together a collection of research papers that provides quantitatively defensible formulations of the influence of military pay, bonuses, retirement benefits, and other factors on enlisted retention.

Author Goldberg, Matthew S.

Title *New Estimates of the Effect of Unemployment on Enlisted Attrition*

Report# ADA172661

PubDate   1985

PubData   Alexandria, VA: Center for Naval Analyses

Subject   military, costs, compensation, attrition

Abstract

This paper provides new estimates of the effect of unemployment on enlisted retention. Unemployment is found to have a positive effect upon the reenlistment rate for seven of the nine rating groups studied, and a positive effect upon both the extension rate and the total retention rate for all nine rating groups. However, the pay elasticities are three to five times as large as the unemployment elasticities, so that decreases in the unemployment rate may be offset by much smaller percentage increases in military pay.

Author   Goldberg, Matthew S., and John T. Warner

Title   "Military Experience, Civilian Experience, and the Earnings of Veterans"

Report#

PubDate   1987

PubData   *Journal of Human Resources*, Vol. 22, Winter 1987, pp. 62–81

Subject   retirement, transitioning, retention, military

Abstract

This paper examines the effects of military experience and civilian experience on the earnings of veterans with the objective of determining the substitutability of these two forms of experience for personnel receiving different types of military training. Analysis of data on 24,000 individuals reveals that more military experience does increase subsequent civilian earnings, but that the relative effect of military and civilian experience varies considerably by military occupational specialty.

Author   Goldberg, Matthew S., and John T. Warner

Title   *Determinants of Navy Reenlistment and Extension Rates*

Report#   CRC 476

PubDate   1982

PubData   Alexandria, VA: Center for Naval Analyses

Subject   retention, personnel flow, compensation, bonuses, Navy, military, development

Abstract

This paper analyzes the effects of regular military compensation and reenlistment bonuses on the probabilities of reenlistment and extension among first-term and second-term Navy enlisted personnel. We find that the pay elasticities vary substantially across occupational categories. For example, the elasticity of the total probability of staying with respect to regular military compensation ranges between 1.12 and 2.72 for first-term personnel and between .94 and 3.78 for second-term personnel. We recommend adopting occupation-specific pay elasticities in forecasting, since the all-Navy pay elasticity will yield misleading results.

Author   Goldich, Robert L.

Title   *Military Retirement and Personnel Management: Should Active Duty Military Careers Be Lengthened?*

Report#   ADA303749

PubDate   1995

PubData   Washington, DC: Library of Congress

Subject   military, performance, cost, retention, transitioning

Abstract

This report discusses whether the current average active duty military career should be lengthened. Proponents argue it could lead to cost savings resulting from more efficient personnel management, and would provide more scope for military career members to obtain more training and experience. Opponents tend to believe that lengthening average careers could result in career retention problems, and could lead to career personnel who were unfit to perform their military duties due to age and consequent lack of physical and mental vigor. Modifications of the current average active duty military career length could thus have substantial implications for the overall defense budget and the military effectiveness of the armed forces. The role of the Congress in these matters is crucial, as overall retirement criteria and retired pay computation formulae for all military members, and detailed personnel management policies for officers, are established by statute. The dominant rationale for shorter careers has been the need to prevent the military effectiveness of the armed forces from being impaired by the presence, on active duty, of people physically incapable—because of age—of performing their military duties. A major secondary rationale for allowing, and requiring, retirement at comparatively earlier ages than most civilian retirement systems is providing a strong career retention incentive.

Author   Gorman, Linda, and G. W. Thomas

Title   "General Intellectual Achievement, Enlistment Intentions, and Racial Representativeness in the U.S. Military"

Report#

PubDate   1993

PubData   *Armed Forces and Society*, Vol. 19, No. 4, pp. 611–624

Subject   military, accession, race, recruit screening, background characteristics, equal opportunity

Abstract

This article examines issues surrounding race, the decision to enlist, and possible strategies for achieving a more representative military force. Previous studies have overstated the role of race in determining the intention to enlist. Race is far less important when general intellectual achievement and poverty are included in the analysis. These findings have an effect on the equity of such strategies as raising AFQT requirements or setting ceilings for black enlistments.

Author   Gray, Rosanna L.

Title   *Influences of High Quality Army Enlistments*

Report#   ADA180562

PubDate   1987

PubData   Monterey, CA: Naval Postgraduate School

Subject   accession, recruiting, personnel flow, high quality personnel, educational benefits, Army, military

Abstract

This thesis investigated the relationship between the quality of soldiers and influences on their enlistment decision. Influences analyzed include economic benefits of enlisting, military advertising, and Army recruiters. Data are from the 1985 New Recruit Survey. The results of principal components and discriminant analysis indicated that educational benefits such as the New GI Bill strongly influenced high quality soldiers. Advertising and recruiters were also important influences on the enlistment decision.

Author   Grissmer, David W., and Judith C. Fernandez

Title   *Meeting Occupational and Total Manpower Requirements at Least Cost: A Nonlinear Programming Approach*

Report#   RAND/P-7123

PubDate   1985

PubData   Santa Monica, CA: RAND

Subject   personnel flow, cost, military, accession, screening, development, retention

Abstract

The Army trains and utilizes enlisted personnel in a wide variety of military occupational specialties, ranging from relatively unskilled to highly skilled, from large specialties to small. Besides developing policies to meet specialty-specific needs, the Army must also develop manpower policies to meet aggregate manpower objectives, including meeting strength goals, observing budget and grade-level limits, and maintaining promotion flows and objective force experience profiles. This paper develops a model designed to evaluate alternative approaches to meeting MOS-specific and aggregate manpower requirements. The analytic model, a cost-minimizing nonlinear programming model, allows the manager to discover the least-cost way of providing the desired quality and quantity profiles of enlisted first-term, second-term, and career personnel in the total force and in each MOS, when MOSs vary as to marginal recruiting costs, training costs, retention histories, and requirements for senior personnel.

Author	Grissmer, David W., Richard L. Eisenman, and William W. Taylor
Title	*Defense Downsizing: An Evaluation of Alternative Voluntary Separation Payments to Military Personnel*
Report#	RAND/ MR-171-OSD/A
PubDate	1995
PubData	Santa Monica, CA: RAND
Subject	transitioning, compensation, personnel flow, military, cost, quality

Abstract

This report documents RAND's research effort on one aspect of the personnel drawdown—how to structure voluntary separation offers to service members to efficiently meet force-reduction objectives. This research was carried out before development of the voluntary separation programs initiated between 1992 and 1994 and was instrumental in shaping them. The authors address the question of what part of the reductions should come from lowered accession levels and what part from increased separations of personnel currently in the service. They identify the criteria that any separation plan should meet and develop a methodology for estimating the acceptance rate of voluntary separation offers. They apply this methodology to evaluate a range of such offers and then address the process of how to structure separation offers to get both the number and type of desired departures as cost effectively as possible. Finally they address questions concerning the financing of such offers by estimating the savings from reduced retirement outlays.

Author	Haber, Sheldon E., Enrique Lamas, and Judith H. Eargle
Title	*A New Approach to Managing the Army Selective Reenlistment Bonus*
Report#	ADA163816
PubDate	1984
PubData	Alexandria, VA: U.S. Army Research Institute for the Behavioral and Social Sciences
Subject	bonuses, SRB, retention, cost, skill needs, personnel flow, Army, military, development, compensation

Abstract

The authors develop a theoretical model of profit maximization in which the Selective Reenlistment Bonus is treated as a wage premium payable to servicemen who are more productive, more costly to recruit and train, and less likely to continue in the Army in the absence of the SRB. Empirical estimation of the model is based on measuring a serviceman's productivity (in terms of his civilian counterpart's occupational wage), recruitment and training costs, and separation rates. The results for E-4 servicemen show that the average SRB for Combat Arms should remain at the current $5,200 level, while Technical occupations should increase to $7,300 and Support Services should drop to $1,900. Overall cost would remain about the same.

Author	Haggstrom, Gus W., Thomas J. Blaschke, Winston K. Chow, and William Lisowski
Title	*The Multiple Option Recruiting Experiment*
Report#	RAND/R-2671-MRAL
PubDate	1981
PubData	Santa Monica, CA: RAND
Subject	accession, recruiting, attrition, retention, bonuses, educational benefits, personnel flow, military

Abstract

Analyzes an experiment, begun in January 1979, to test the effectiveness of new enlistment incentives aimed primarily at high-quality males for hard-to-fill occupational specialties. The incentives included a two-year enlistment option, enhanced post-service educational benefits, and an "IRR option" permitting recruits to choose between reserve and active duty after completing initial training. The enlistment responses to the options were disappointing: None of the options elicited a sizable response. Only the IRR option showed promise as an incentive for combat arms enlistees. An examination of the policy issues associated with the incentives suggests that shorter-term enlistments and educational benefits may even be detrimental to the services in the long run because they lead to lower retention at the end of the first term of service.

Author	Hanser, Lawrence M., Joyce N. Davidson, and Cathleen Stasz
Title	*Who Should Train? Substituting Civilian-Provided Training for Military Training*
Report#	RAND/R-4119-FMP
PubDate	1991
PubData	Santa Monica, CA: RAND
Subject	training, civilian substitution, accession, cost, military, development

Abstract

The initial skill training (IST) of military enlisted personnel has historically been conducted by the military services. In light of expected changes in the size and structure of the force, and the increasing importance of the reserve forces, Congress has asked whether initial skill training for technical occupations could be provided by civilian institutions. This report describes an analysis of the issues associated with the feasibility of using civilian institutions for this purpose. There is sufficient evidence that civilian organizations can provide military technical training; the more important question is how to choose from among the alternatives. For evaluating training options, the authors developed a conceptual framework based on selecting the lowest-cost training scenario that produces a given level of trained man-years. They conclude that (1) many military occupations are amenable to civilian training, (2) former and existing programs have not been adequately evaluated, (3) civilian-provided IST appears to have benefits in some circumstances, and (4) there are institutional barriers to implementation. They recommend the development of a joint-service working group on training policy and the inauguration of a series of demonstration projects.

Author	Harman, Joan
Title	*Three Years of Evaluation of the Army's Basic Skills Education Program*
Report#	ADA170476
PubDate	1984
PubData	Alexandria, VA: U.S. Army Research Institute for the Behavioral and Social Sciences
Subject	training, development, performance, military, accession

Abstract

This report summarizes three years of evaluation research on the Army's Basic Skills Education Program. It describes analysis of standard, pilot, and revised programs as well as programs under development. Overall findings reveal that all programs tend to move participants in the direction of Army goals for basic skill improvement, although a substantial number of soldiers fail to meet program criteria. Army decisionmakers can be guided by these findings in planning the future of the Basic Skills Education Program.

Author	Hawes, Eric A.
Title	*An Application of Survival Analysis Methods to the Study of Marine Enlisted Attrition*
Report#	ADA226565
PubDate	1990

PubData   Monterey, CA: Naval Postgraduate School

Subject   military, Marines, attrition, screening, quality, background characteristics, accessing

Abstract

This thesis is an application of survival analysis methods to study first-term enlisted attrition from the Marine Corps. The data comprise over 99 percent of all enlisted accessions into the Marine Corps between October 1983 and August 1988. The majority of the findings concerning the effects of the covariates on attrition are consistent with published results from previous military attrition studies. Two findings of the thesis are new. First, the attrition behavior of alternative high school credential holders varied significantly according to credential type. Second, the relationship between aptitude and attrition behavior appears to have weakened in recent years.

Author   Hogan, Paul F., D. Alton Smith, and Stephen D. Sylvester

Title   "The Army College Fund: Effects on Attrition, Reenlistment, and Cost"

Report#

PubDate   1991

PubData   Alexandria, VA: U.S. Army Research Institute for the Behavioral and Social Sciences (in *Military Compensation and Personnel Retention*, edited by C. L. Gilroy, D. K. Horne, and D. A. Smith)

Subject   military, Army, educational benefits, recruiting, attrition, retention, costs

Abstract

This paper analyzes the effects of the Army College Fund—a special Army-specific educational benefit—on attrition, reenlistment, and costs, using data from a 10 percent sample of the Army FY 1982 recruit cohort. The results suggest that ACF benefits have a small, but statistically insignificant effect on the probability that a recruit will remain in service over his initial term. The benefits do, however, reduce the probability that the recruit will choose to reenlist.

Author   Horne, David C., and Mary Weltin

Title   *Determinants of Army Career Intentions*

Report#   ADA178672

PubDate   1985

PubData   Alexandria, VA: U.S. Army Research Institute for the Behavioral and Social Sciences

Subject   accession, recruiting, personnel flow, recruit screening, background characteristics, leadership, recruit screening, Army, military

Abstract

This research models the career intentions and choice of tour length of U.S. Army recruits. We find that the reasons for enlisting can be used to create four basic motivational factors using factor analysis. These factors, in addition to a number of demographic and socioeconomic variables, appear to be significant determinants of career intentions and tour length. The results of this study demonstrate that the factors that motivate citizen soldiers appear to differ from the factors that motivate career soldiers. The Army may be able to use this information to design incentive programs and advertising strategies to attract various groups of potential recruits.

Author   Hosek, James R., and Christine E. Peterson

Title   *Reenlistment Bonuses and Retention Behavior: Executive Summary*

Report#   RAND/R-3199/1-MIL

PubDate   1985

PubData   Santa Monica, CA: RAND

Subject   compensation, bonuses, retention, quality, flexibility, recruit supply, personnel flow, military, development, cost

Abstract

This report, an executive summary of RAND Report R-3199-MIL, presents a non-technical discussion of the most policy-relevant findings of research on the effects of bonuses on retention behavior. The findings suggest that, overall, the reenlistment bonus program should be continued and perhaps expanded. It enables the services to respond quickly to changes both in labor supply, such as those created by economic and demographic cycles, and in labor demand, such as those created by changes in weapons systems or force deployment. Bonuses are effective in increasing retention rates and promoting longer terms of service. Since they are not part of base pay, they do not directly increase the potential retirement outlays as an increase in base pay would. Their power and flexibility make them a valuable aid in managing the size and shape of the career force.

Author	Hosek, James R., Christine E. Peterson, and Joanna Zorn Heilbrunn
Title	*Military Pay Gaps and Caps*
Report#	RAND/MR-368-P&R
PubDate	1994
PubData	Santa Monica, CA: RAND
Subject	compensation, bonuses, costs, military, quality, retention, personnel flow, development

Abstract

This report investigates the military/civilian pay gap and its implications for capping military pay increases. The pay gap is defined as the percentage difference in military versus civilian pay growth as measured from a given starting point. The index currently used for civilian pay growth is the Employment Cost Index (ECI), which reflects pay growth in the civilian labor force at large. The authors instead recommend measuring civilian pay growth for the subset of civilian workers whose composition by age, education, occupation, gender, and race/ethnicity represents that of active duty military personnel. The authors do so via construction of a Defense Employment Cost Index (DECI). They compare pay gaps based on the ECI vs. the DECI, and present DECI-based pay gaps for officer and enlisted personnel by gender and seniority and for occupational and age categories. The authors then consider the implications of these pay gaps for capping military pay.

Author   Hosek, James R., Christine E. Peterson, Jeannette VanWinkle, and Hui Wang

Title   *A Civilian Wage Index for Defense Manpower*

Report#   RAND/R-4190-FMP

PubDate   1992

PubData   Santa Monica, CA: RAND

Subject   compensation, military, development, cost

Abstract

Current estimates comparing wage growth in the military and civilian sectors suggest that military basic pay grew 11.8 percent slower than civilian wages from FY 1982 to FY 1991. Yet over the same period recruit quality and retention showed practically no sign of deterioration. The authors developed a new civilian wage index (the DECI) for military personnel designed to represent the active duty military population and its civilian wage opportunities, and found only a 4.7 percent deficit over that period. DECI relative wage growth patterns track accession and retention trends over the period better than the currently used index, the ECI. The DECI also shows how the military/civilian pay gap varies among different subgroups in the military—young, high school only graduates show parity while older, more educated personnel show a deficit; enlisted personnel show parity while officer personnel show a deficit. Subgroup comparisons demonstrate the DECI's potential to provide specific information to those who manage and set policy for military personnel.

Author   Hosek, James R., John Antel, and Christine E. Peterson

Title   *Who Stays, Who Leaves? Attrition Among First-Term Enlistees*

Report#   RAND/N-2967-FMP

PubDate   1989

PubData   Santa Monica, CA: RAND

Subject   attrition, accession, military, recruit characteristics

Abstract

For nearly a decade, 35-month attrition rates in the volunteer armed forces have exceeded 25 percent. This Note (reprinted from *Armed Forces and Society*, Vol. 15, No. 3, Spring 1989) identifies the determinants of attrition behavior of high school students and graduates—the population groups most significant for recruiting today's higher-quality force. To improve the capability to predict who is likely to stay and who is likely to leave, and to understand why attrition occurs, the authors employ a unique microdata set and specify a statistical model that analyzes male attrition behavior jointly with enlistment. The results suggest that attrition is higher among people who enter rashly and without firm career goals, who have a history of employment instability, and who do not expect to obtain further education. Policy implications are discussed.

Author   Hosek, James R., Richard L. Fernandez, and David W. Grissmer

Title   *Active Enlisted Supply: Prospects and Policy Options*

Report#   RAND/P-6967

PubDate   1984

PubData   Santa Monica, CA: RAND

Subject   personnel flow, accession, compensation, attrition, retention, recruiting, military, Army

Abstract

In this paper the authors show how the forces shaping enlisted supply in the 1980s will affect the services' abilities to attract and retain the numbers and types of individuals they want. They give special attention to the Army because its past recruiting problems have sparked greatest concern. For both the Department of Defense (all services together) and the Army, they present forecasts of high-quality male enlistments, of first- and second-term retention rates, and of the enlisted force structure. They also examine some alternative policy options that could be used if the predicted enlistment levels, retention rates, and force structures deviate from those desired. Finally, they indicate areas in which further analysis would be useful.

Author	Hunter, Cardell S.
Title	*Measuring the Impact of Military Family Programs on the Army*
Report#	ADA182778
PubDate	1987
PubData	Carlisle Barricks, PA: U.S. Army War College
Subject	military, family support, performance, retention, morale, child care, Army, development

Abstract

This report researches the effectiveness of the Army's Military Family Programs and attempts to measure their effect on unit readiness, family wellness, and soldier retention. In recent years the Army has come to recognize the importance of the Army family and to acknowledge its obligation to family support. Toward that end, programs such as Family Advocacy, Family Child Care, Sponsorship, Exceptional Family Member Program, and the Army Community Service now receive the attention of the Army. The basic question is whether or not the Army is getting its money's worth from these programs. Data were gathered using a literature search, the development and employment of a questionnaire, and personal interviews with Army officers at the United States Army War College. The research for this project unequivocally indicated that Military Family Programs are on the whole very positive and do serve to enhance family wellness, unit readiness, and soldier retention.

Author	Kawahara, Yoshito, L. A. Palinkas, R. Burr, and P. Coben
Title	*Suicides in Active-Duty Enlisted Navy Personnel*
Report#	ADA219287
PubDate	1990
PubData	San Diego, CA: Naval Health Research Center

Subject   military, Navy, health, background characteristics

Abstract

Three hundred seventy-three cases of completed suicide among Navy enlisted personnel occurring during 1974 to 1985 were identified from a Naval Health Research Center data file. The suicide rate of 6.59 per 100,000 person-years was lower than the overall U.S. rate. Trends among age groups, occupations, and gender are examined.

Author   Kearl, Cyril E., and Abraham Nelson

Title   "The Army's Delayed Entry Program"

Report#

PubDate   1992

PubData   *Armed Forces and Society*, Vol. 18, No. 2, pp. 253–268

Subject   military, recruiting, accession, DEP, background
            characteristics, enlistment bonuses

Abstract

This article examines the factors contributing to Delayed Entry Program attrition, focusing on DEP loss for the U.S. Army in 1986 and 1987. Personal characteristics were found to be important, including age, gender, race, dependent status, and high school status. Other factors that increased DEP loss were improving economic conditions and reduced recruiting incentives. Enlistment benefits such as bonuses and Army job training helped reduce attrition.

Author   Kim, Choongsoo

Title   *The All-Volunteer Force: An Analysis of Youth
           Participation, Attrition, and Reenlistment*

Report# ADA185416

PubDate 1982a

PubData Ohio State University: Center for Human Resource
           Research

Subject   recruiting, accession, background characteristics, military, job satisfaction

Abstract

Major findings from the initial interview of the 1979 National Longitudinal Survey of Youth are summarized. The armed forces contain 6.7 percent of males and 0.6 percent of females in the 18–21 cohort; however, there are differences in the participation rate of different demographic and social groupings based on race, educational achievement and expectations, marital status, and professional background of families. The armed forces are enlisting young men with backgrounds and abilities comparable to those youth who are employed full-time in the civilian labor market and young women with backgrounds and abilities that are higher than their civilian full-time employed counterparts. Comparison of 18 measures of different job aspects clearly shows that armed forces personnel are less satisfied than their civilian counterparts. Women in the military see more favorable job aspects than men, but are still less satisfied than their civilian counterparts. Male military personnel are paid 12 percent less and females are paid 18 percent more than their civilian counterparts; however, values of enlistment bonuses, educational benefits, and other military benefits were not included in the calculations. About 1.2 million males and 0.6 million females aged 18 to 21 are interested in serving in the armed forces and about 25 percent of men and 38 percent of women in the services expressed positive reenlistment intentions. School attendance is the single greatest identifiable reason given for not enlisting by those who had talked with a recruiter but did not join.

Author   Kim, Choongsoo

Title   *Youth and the Military Service: 1980 National Longitudinal Survey Studies of Enlistment, Intentions to Serve, Reenlistment, and Labor Market Experience of Veterans and Attriters*

Report#   ADA185410

PubDate   1982b

PubData   Ohio State University: Center for Human Resource Research

Subject accession, recruiting, attrition, retention, recruit characteristics, military, job satisfaction

Abstract

Five studies of military manpower issues based on data from the second survey of a cohort of youth aged 14 to 21 on January 1, 1979, and part of the National Longitudinal Surveys of Labor Force Experience (NLS), are provided. Studies focus on characteristics of participants in the armed forces, characteristics of enlistees, factors in enlistment decisions, reenlistment, and post-military labor market experiences. The socioeconomic status and quality of survey respondents in the armed forces were found to be similar to those of civilian youth employed full time. Interservice comparisons, however, indicated disparities among the four services. Comparison of 1979 enlistees with those who enlisted in 1978 revealed declines in parental education attainment, proportion completing high school, and mean AFQT scores. Long-run returns were cited as the chief reason for enlisting with training opportunities being the most often cited long-run return. The majority of youth felt serving in the military was definitely or probably a good thing, but few said they would try to enlist in the future. Among those who talked to recruiters, going to school was frequently cited as the reason for those not enlisting. Job satisfaction was an important factor affecting reenlistment decisions, followed by marital status and the presence of a child. Among male attriters, mean AFQT scores, college enrollment rates, and weekly earnings were lower, and unemployment rates were higher than for those who remained in the armed forces. Among females, mean AFQT scores and unemployment rates were higher, college enrollment rates were similar, and weekly earnings were substantially higher than female veterans.

Author Kim, Choongsoo, Gilbert Nestel, Robert L. Phillips, and Michael Borus

Title *The All Volunteer Force: 1979 NLS Studies of Enlistment, Intentions to Serve, and Intentions to Reenlist*

Report# ADA185413

PubDate 1980

PubData   Ohio State University: Center for Human Resource
Research

Subject   accession, recruiting, recruit characteristics, attrition,
retention, personnel flow, military

Abstract

Three studies of the All-Volunteer Force system based on data from
the first survey of a cohort of youth aged 14 to 21 on January 1, 1979,
and part of the National Longitudinal Surveys of Labor Force
Experience (NLS), are provided. Studies focus on factors affecting
enlistment, intentions to serve, and intentions to reenlist. Among recent male high school graduates, military service is favored over college and other civilian pursuits as local labor market conditions de-
teriorate; individuals who desire to complete higher education are
most likely to enter college; and those who want occupational training are more likely to enlist than to remain in the civilian sector.
Among 14 to 17 year olds, excluding high school seniors, positive intentions to serve are inversely related to educational attainment and
socioeconomic status and positively correlated with perception of
approval. Among 18 to 21 year olds, excluding high school seniors,
intentions to enlist are positively related to post-service educational
benefits among blacks and Hispanics and occupational training
among whites. Among white high school seniors, intentions to serve
are inversely related to socioeconomic status and positively related
to the desire to obtain occupational training and poorer labor market
conditions. Job satisfaction, marriage, and length of service are related to reenlistment decisions.

Author   Kirby, Mary A.

Title   *A Multivariate Analysis of the Effects of the VSI/SSB
Separation Program on Navy Enlisted Personnel*

Report#   ADA265446

PubDate   1993

PubData   Monterey, CA: Naval Postgraduate School

Subject   military, Navy, transitioning, bonuses, compensation

Abstract

This thesis investigates the behavior of Navy enlisted personnel who were eligible for, and offered, early voluntary separation under one of two monetary incentive programs during FY 1992. The two programs were the voluntary separation incentive (VSI) and the special separation bonus (SSB). Multivariate logit models were estimated to explain the decision to accept a voluntary separation incentive and the decision of which program to accept. The results show that the statistically significant factors related to the separation decision are consistent with simple economic theory.

Author	Kirby, Sheila Nataraj, and Harry J. Thie
Title	*Enlisted Personnel Management: A Historical Perspective*
Report#	RAND/MR-755-OSD
PubDate	1996
PubData	Santa Monica, CA: RAND
Subject	military, accession, development, transitioning, promoting, compensation

Abstract

Enlisted force management is concerned with meeting national military manpower requirements with the nation's citizens. It attempts to balance the demand for enlisted personnel, as determined by the requirements process, with the supply of enlisted personnel, in a cost-effective manner. This report presents a historical view of enlisted force management from two perspectives. First, it provides a chronological account of how external influences have shaped enlisted force management and the evolution of enlisted management practices. Second, it describes some recurring themes that run through the history of the enlisted force: quality, specialization, integration, separation, and compensation.

Author	Klein, Stephen, Jennifer Hawes-Dawson, and Thomas Martin
Title	*Why Recruits Separate Early*

Report#   RAND/R-3980-FMP

PubDate   1991

PubData   Santa Monica, CA: RAND

Subject   attrition, accession, recruit screening, personnel flow, military, background characteristics

Abstract

Approximately 27 percent of military recruits are discharged before the end of their first term for reasons that result in an adverse Interservice Discharge Code (ISC). ISCs indicate the justification for the discharge, but not necessarily what the recruit did to deserve the separation. The ISC system also lacks a way to indicate multiple reasons for a discharge. This study investigated the actual reasons for early, adverse separations through an analysis of recruits' hard-copy personnel records. It also examined the relationship between these reasons and such recruit characteristics as gender, race, service, year of entry, education, and military occupational specialty. The most prevalent reasons for early discharge involved work/duty, training, minor offense, and mental health problems. Most of the recruits separated for three or more reasons. Certain causes—major and minor criminal offenses, drugs, and alcohol—tended to occur together. Recruits who had one or more of these four problems were unlikely to separate for mental health reasons. Recruits who separated because of homosexuality were unlikely to have work/duty problems. The results were fairly consistent across services.

Author   Lakhani, Hyder, Shelley Thomas, Jeffrey Anderson, Curtis Gilroy, and Cavan Capps

Title   *Army European Tour Extension: An Interdisciplinary Approach*

Report#   ADA173525

PubDate   1985

PubData   Alexandria, VA: U.S. Army Research Institute for the Behavioral and Social Sciences

Subject   military, costs, bonuses, compensation, retention

Abstract

An objective of this paper is to analyze the effect of economic and non-economic factors on the decisions to extend the European tours of servicemen by 12 months. The research used the Army Research Institute's 1983 survey of Army families in Europe. Employing multivariate techniques, the results revealed that the extension probabilities increased with an increase in the amounts of the proposed extension bonuses as well as with increase in satisfaction with job and family life. Smaller lump-sum bonuses were determined cost-effective compared to higher amounts of monthly installment bonuses as incentives for extension. Both the lump sum as well as the installment bonuses were, however, cost-effective compared to the Permenent Change of Station (PSC) costs.

Author   Laurence, Janice H., Jennifer Naughton, and Dickie A. Harris

Title   *Attrition Revisited: Identifying the Problem and Its Solutions*

Report#   ADA309172

PubDate   1996

PubData   Alexandria, VA: U.S. Army Research Institute for the Behavioral and Social Sciences

Subject   military, attrition, accession, screening

Abstract

This report highlights the known and suggested causes, correlations, and contributors to first-term attrition. Personal characteristics such as education credential, aptitude, gender, and age are discussed first followed by a description of organizational and situational influences. Better coordinated multivariate selection approaches to attrition are suggested as are realistic previews of military life. A pivotal factor in attrition is its management/policy control at various levels. This aspect of attrition must be understood before other reduction strategies are introduced.

Author   Lewis, Philip M.

Title   *Family Factors and the Career Intent of Air Force Enlisted Personnel*

Report#   ADA164899

PubDate   1985

PubData   Maxwell Air Force Base, AL: Leadership and Management Development Center

Subject   retention, family support, job satisfaction, personnel flow, Air Force, military, development

Abstract

The effect of spouse attitudes and attributes on the career intent and job-related attitudes of Air Force enlisted personnel was assessed using the Air Force's new Family Survey (AFFS) to measure spouse attitudes and the Organizational Assessment Package to assess the Air Force member's career intent and job attitudes. The factor structure of the AFFS confirmed its potential utility for assessing critical family variables. It proved possible to predict the career intent and job satisfaction of Air Force members from spouse attitudes and other family variables, most importantly from the compatibility of the marital pair's work schedules, the positiveness of the spouse's view of the Air Force and, for career intent only, the perceived stressfulness of the Air Force member's job and Air Force life for the family.

Author   Lofink, Diane L. H.

Title   *The Effect of Providing On-Site Child Care on Personnel Productivity, Morale, and Retention*

Report#   ADA237494

PubDate   1990

PubData   Monterey, CA: Naval Postgraduate School

Subject   compensation, retention, performance, child care, military, Navy, development

Abstract

This thesis investigates the possible effect of on-site child development centers on the productivity, morale, and retention of Naval officers and enlisted personnel. A written survey was conducted of active duty Navy personnel with dependents under age 13, assigned to eight Navy shore installations, four of which offer child care and four of which do not. Statistical analyses conducted while controlling for other factors suggest that on-site centers do not significantly increase or decrease the probability of either work interference or career influence.

Author	Mael, Fred A., and Blake E. Ashforth
Title	"Loyal from Day One: Biodata, Organizational Identification, and Turnover Among Newcomers"
Report#	
PubDate	1995
PubData	*Personnel Psychology*, Vol. 48, pp. 309–333
Subject	recruiting, accession, background characteristics, attrition, recruit screening, personnel flow, military, Army

Abstract

This paper attempts to use biodata to uncover behavioral and experiential antecedents of organizational identification (OID), and to demonstrate one way in which theory can be used in the development and analysis of objective biodata. The biodata correlates of organizational identification were assessed with a sample of 2,535 new U.S. Army recruits. Four biodata factors emerged: activities involving outdoor work or pastimes, general lifestyle and socialization, a preference for group attachments, and diligent involvement in intellectual pastimes. Results show that both OID and the biodata predicted subsequent attrition across periods from 6 to 24 months.

Author	Manganaris, Alex G., and Chester E. Phillips
Title	*The Delayed Entry Program: A Policy Analysis*

Report#   ADA167847

PubDate   1985

PubData   Alexandria, VA: U.S. Army Research Institute for the
Behavioral and Social Sciences

Subject   accession, recruiting, attrition, cost, DEP, military, Army,
personnel flow

Abstract

This research explores the effect of the Delayed Entry Program on preaccession and first-term behavior of Army enlisted personnel. Two models, one exploring the DEP and preaccession behavior (DEP loss) and the other examining the effect of the DEP on first-term attrition, are combined to explore DEP policy tradeoffs. Recruiting costs and training costs are used as a measure of DEP effect on force management. Specific Military Occupational Specialties are examined and marginal cost measures are developed. Results show that when cost is considered, longer DEP lengths are recommended, when possible, for almost all personnel. Findings show that marginal costs vary by personnel characteristics and MOS job assignment.

Author   Manganaris, Alex G., and Edward J. Schmitz

Title   *Impact of Delayed Entry Program Participation on First
Term Attrition*

Report#   ADA178669

PubDate   1985

PubData   Alexandria, VA: U.S. Army Research Institute for the
Behavioral and Social Sciences

Subject   attrition, accession, personnel flow, DEP, background
characteristics, military, Army

Abstract

This research analyzes the effect of the Delayed Entry Program participation on attrition in the first term. Using logistic regression, equations were estimated to determine the probability of attrition in 13 Army jobs. Independent variables used in this analysis were education, time in DEP, gender, race/ethnic background, and Armed

Forces Qualification Test (AFQT) score. Results showed that in 9 of 13 equations the DEP had a significant effect on the probability of attrition. Results consistently showed that the longer individuals participate in the DEP, the less likely they are to attrit in their first term.

Author	Manganaris, Alex G., and Edward J. Schmitz
Title	*Projecting Attrition by Military Occupational Specialty*
Report#	ADA168183
PubDate	1984
PubData	Alexandria, VA: U.S. Army Research Institute for the Behavioral and Social Sciences
Subject	attrition, accession, job assignment, background characteristics, personnel flow, cost, Army, military

Abstract

This research analyses the attrition rates of various enlisted personnel groups within different Army job assignments. Three regression equations are developed to project the attrition rate of eight demographic groups to 76 Military Occupational Specialties. Education, sex, AFQT, along with MOS assignment are the independent variables. The rates generated by these equations show where important tradeoffs exist with respect to personnel allocation and the expected rate of attrition. The effect of AFQT scores on attrition varies by MOS, with some MOSs showing more sensitivity to differences in AFQT scores. High school graduates and those with GEDs have lower attrition than nongraduates. Women have higher attrition than men and blacks have significantly lower attrition than whites or Hispanics.

Author	Marcus, Alan J.
Title	*Summary Report: Manning the 600-Ship Navy*
Report#	CRM 85-111.10
PubDate	1985
PubData	Alexandria, VA: Center for Naval Analyses

Subject   Navy, personnel flow, accession, performance, cost, compensation, promotion, military, quality

Abstract

This study summarizes a series of projects designed to improve the Navy's ability to set manpower requirements and to develop cost-effective compensation polices to fill these requirements. The analyses included several efforts to improve the methodology used to define manpower requirements. A computer model was designed to help in the analysis of the effect of changes in the size of the fleet on requirements at the individual billet level. Development of methodologies to assess the potential for civilian substitution and to define test score and educational requirements for accessions was also completed. The effect of personnel characteristics and Navy training on the performance of enlisted personnel was the subject of two separate research efforts. Finally, the effects of compensation policy on high-quality personnel and of sea pay on hard-to-fill sea-intensive billets were the subjects of two studies of retention behavior.

Author   Marshall, Ernest V.

Title   *The Career Job Reservation System. Is the Issuance System Best Serving the Needs of the Air Force?*

Report#   ADA241065

PubDate   1990

PubData   Maxwell AFB, AL: Air War College

Subject   military, bonuses, compensation, development, retention, transitioning

Abstract

The Career Job Reservation issuance system, a part of the Selective Reenlistment Bonus Program, functions on a quota basis and is perceived by commanders and airmen to have many inequities. Since the Air Force must restrict the size of the career enlisted force, reenlistment is permitted only if a valid requirement can be filled. To prevent shortages and surpluses, monthly reenlistment quotas are established and controlled by the use of a reservation system. Although the system is serving the basic needs of the Air Force, major changes are in order to remove some of the inequities, correct perceptions, and improve procedures. The proposed changes involve (1) converting from use of protected months to protected quarters, (2) realigning the priority of quality factors, (3) rank-ordering the waiting list, (4) expanding the eligibility for wing commander overrides, (5) improving publicity and advertising to airmen and commanders, and (6) developing a subjective value assessment score to be given by the unit commander at the time of selection for reenlistment.

Author   Mickleson, William T., and C. Peter Rydell

Title   *Aggregate Dynamic Analysis Model (ADAM) for Air Force Enlisted Personnel: User's Guide*

Report#   RAND/N-3020/1-AF

PubDate   1989

PubData   Santa Monica, CA: RAND

Subject   personnel flow, attrition, retention, promotion, military, development, retention

Abstract

The Aggregate Dynamic Analysis Model (ADAM) projects active duty Air Force enlisted personnel and their budget costs that will result from user-specified management actions and background economic conditions for 12 years into the future. These projections are for the aggregate force (total across all specialties) by pay grade, years of service, and category of enlistment. The management actions include accessions, reenlistment bonuses, early releases, early reenlistments, promotions, and involuntary separations. The background economic conditions controlled for by the model include civilian unemployment rate, ratio of military wages to civilian wages, and the Consumer Price Index. This user's guide volume describes what the model does, explains how to run the model, and gives example input and output screens. A companion volume, N-3020/2, presents technical documentation for ADAM.

Author   Moore, S. Craig, J. A. Stockfisch, Matthew S. Goldberg, Suzanne M. Holroyd, and Gregory G. Hildebrandt

Title   *Measuring Military Readiness and Sustainability*

Report#   RAND/R-3842-DAG

PubDate   1991

PubData   Santa Monica, CA: RAND

Subject   readiness, personnel requirements, measurement,
             flexibility, training, military, performance

Abstract

This report reviews the state of the art in readiness and sustainability measurement and develops a strategic concept design for measurements that would better serve high-level defense decisionmakers. The authors identify (1) incremental improvements that would raise the value of information derived from current reporting and analysis systems and (2) a new concept for assessing readiness and sustainability that would integrate several existing reporting and analysis approaches. The findings indicate that today's indicators of readiness and sustainability do not provide high-level defense decisionmakers with appropriate information. Estimates of the levels of activity that U.S. forces could achieve over time in different contingencies would be more useful. Using continuous numerical scales and showing changes during a contingency, such integrated assessments should prove more sensitive to resource-level changes and allow easier comparisons from year to year.

Author   Norrblom, E. M.

Title   *The Returns to Military and Civilian Training*

Report#   RAND/R-1900-ARPA

PubDate   1976

PubData   Santa Monica, CA: RAND

Subject   training, development, military, attrition, retention

Abstract

This report is an examination of the economic effects of formal military vocational training and informal on-the-job training acquired while working in a military specialty. The findings support the economic and statistical significance of military training in explaining differences in the post-service wages of separatees. Formal vocational training in the military tends to have a significantly positive effect on post-service wages if individuals enter civilian occupations related to their military specialties. However, informal on-the-job training in military specialties parallel to the current civilian occupations of separatees does not have a significantly positive effect on their post-service opportunities. The report also evaluates the extent to which the returns to various types of training were overestimated or underestimated by previous studies. It shows that the returns are significantly smaller to academic training, but greater to civilian on-the-job training, than indicated by past studies.

Author   OASD (FM&P)

Title   *Military Women in the Department of Defense*

Report#   ADA206373

PubDate   1984

PubData   Washington, DC: OASD

Subject   gender, personnel flow, force structure, military

Abstract

This report presents summary data on women and their status in all components of the armed forces. It makes selected demographic comparisons between men and women as a means of providing an overview.

Author   Orvis, Bruce R.

Title   *Relationship of Enlistment Intentions to Enlistment in Active Duty Services*

Report#   RAND/N-2411-FMP

PubDate   1986

PubData   Santa Monica, CA: RAND

Subject   recruiting, recruit characteristics, accession, military, personnel flow, gender

Abstract

This Note presents work on the relationship between enlistment intention information and active duty enlistments. Earlier RAND research demonstrated a significant relationship between nonprior service respondents' stated enlistment intentions in the Youth Attitude Tracking Study (YATS) and their actual subsequent enlistment actions. Since women were not included in the YATS initially, the research was based on results for nonprior service men. This Note highlights the men's results and reports and compares results for female and male respondents in recent YATS waves. The results indicate that enlistment intention information is useful for both sexes. However, they suggest it is probably less helpful for women than for men. The results also indicate that people stating negative enlistment intentions are an important source of enlistees, and that simple comparisons of positive intention rates for the two sexes can overstate women's interest in military service.

Author	Orvis, Bruce R., and Martin T. Gahart
Title	*Enlistment Among Applicants for Military Service: Determinants and Incentives*
Report#	RAND/R-3359-FMP
PubDate	1990
PubData	Santa Monica, CA: RAND
Subject	accession, recruiting, background characteristics, bonuses, educational benefits, military, personnel flow, quality

Abstract

Drawing primarily on results from the 1983 Survey of Military Applicants, a survey of male youths without prior military service who took the written test to qualify for the military, this report presents findings on three research issues: (1) the implementation of the programs associated with the Enlistment Bonus Test, a nationwide experiment on the effectiveness of various cash enlistment bonuses; (2) factors that lead nonprior service applicants to enlist;

and (3) the development and evaluation of methods of using survey enlistment intention responses to questions about the likelihood of enlisting under specific hypothetical options to help predict the effect of implementing these options. The results support the implementation of the test and, thus, the viability of its conclusions; indicate that civilian job opportunities, social support for enlisting, college plans, and finances affect the enlistment decisions of applicants, suggesting that messages emphasizing the job stability, training, and educational opportunities provided by the military could be effective in advertising and recruiting efforts; and validate the use of intention information in the design and execution of tests of prospective enlistment options.

Author	Orvis, Bruce R., and Susan Way-Smith
Title	"Reducing the Army's NCO Content: Estimated Cost Savings"
Report#	RAND
PubDate	unpublished
PubData	Santa Monica, CA: RAND
Subject	personnel flow, promotion, military, Army, cost, development

Abstract

This draft reports results from a special, short-term analysis carried out to assess the potential cost savings resulting from a more junior Army enlisted force, in which 10 percent of the post-drawdown Army NCO strength would be shifted to grades E-1 to E-4. The proposal was developed by the Army as part of its effort to consider methods of reducing MPA expenditures. The results are based on a number of assumptions and estimates; they thus convey information about the general magnitude of the various costs and savings discussed, rather than precise dollar values. Estimated savings for the more junior force option range from $146 to $169 million per year and would be realized only after a four-year phase-in period in which transition expenses would offset savings.

Author   Orvis, Bruce R., James R. Hosek, and Michael G.
            Mattock

Title   *PACER SHARE Productivity and Personnel Management
            Demonstration: Third Year Evaluation*

Report#   RAND/MR-310-P&R

PubDate   1993

PubData   Santa Monica, CA: RAND

Subject   performance, employee involvement, work systems,
            cost, compensation, promotion, retention, transitioning,
            development

Abstract

This report describes the PACER SHARE Productivity and Personnel
Management Demonstration and the plan that has been developed
to evaluate it. The report also presents statistical results concerning
the quality of work life, organizational flexibility, work quality, and
cost savings during the baseline period prior to the demonstration
and throughout the demonstration's first three years. PACER SHARE
is a five-year demonstration being conducted at the Directorate of
Distribution (DS) within the Sacramento Air Logistics Center (SM-
ALC) under the legal authority of the Office of Personnel Management. Its purpose is to determine whether several changes in
federal civil service practices being tried on an experimental basis
will improve organizational productivity, flexibility, and quality of
work life, while sustaining (or improving) the quality and timeliness
of work and the capability to mobilize during emergency or wartime.
The DSs at the four remaining ALCs (which perform similar functions) serve as the comparison sites. The demonstration formally
began in February 1988 after several years of planning. If effective,
the interventions will subsequently be considered for wider application.

Author   Orvis, Bruce R., Martin T. Gahart, Alvin K. Ludwig, and
            Karl F. Schutz

Title   *Validity and Usefulness of Enlistment Intention
            Information*

Report#   RAND/R-3775-FMP

PubDate   1992

PubData   Santa Monica, CA: RAND

Subject   recruiting, background characteristics, military, accession, personnel flow

Abstract

In synthesizing the results of several years' analysis, this study reports on the validity and application of enlistment intention information for nonprior service youth (i.e., those who have not previously served in the military). The analyses demonstrate that intentions predict enlistment, and suggest that they augment information about a person's likelihood of enlisting beyond what can be discerned from his background characteristics. Aggregate intention levels are significantly related to regional high-quality enlistment rates. Survey data can predict the effects of certain enlistment options relative to others and can approximate the results of field tests. Finally, the report indicates that the negative intention group is an important source of enlistees, and describes conditions that strengthen and weaken the relationship between stated intentions and enlistment actions.

Author   Orvis, Bruce R., Michael T. Childress, and J. Michael Polich

Title   *Effect of Personnel Quality on the Performance of Patriot Air Defense System Operators*

Report#   RAND/R-3901-A

PubDate   1992

PubData   Santa Monica, CA: RAND

Subject   performance, training, quality, Army, military, AFQT, accession

Abstract

This report examines the linkage between the quality of enlisted personnel (in terms of aptitude score) and their ability to operate the Patriot air defense missile system. The intent was to help the Army set appropriate performance standards and estimate the effects of personnel quality on operational performance. The study finds that the Armed Forces Qualification Test score has a direct, consistent effect on the outcomes of air battles, both in terms of knowledge assessed by written tests and in actual performance simulations. Specifically, soldiers with higher AFQT scores can be expected to suffer significantly less asset damage, destroy more hostile aircraft, and be more effective in missile conservation. The study also finds that a one-level change in AFQT category equaled or surpassed the effect of a year of operator experience or of frequent training, a finding that has significant readiness and cost implications, since higher-quality soldiers require less training and operator experience. Finally, the study finds that next to AFQT, operator and unit experience are the factors that most consistently affect performance.

Author   Orvis, Bruce R., Narayan Sastry, and Laurie L. McDonald

Title   *Military Recruiting Outlook: Recent Trends in Enlistment Propensity and Conversion of Potential Enlisted Supply*

Report#   RAND/MR-677-A/OSD

PubDate   1996

PubData   Santa Monica, CA: RAND

Subject   military, Army, recruiting, accession, quality, background characteristics

Abstract

This report describes recruiting trends through early 1995, focusing on changes in youth enlistment propensity and the Army's ability to "convert" the potential supply of recruits into actual enlistments. Using updated survey data and methods of analyzing propensity, it concludes that the potential supply of recruits remains higher in FY95 than it was during 1989, when recruiting results were good. However, the latest survey results indicate some downturn in youth interest in military service. When that downturn is coupled with the large increase in accession requirements during FY96 and FY97, the ratio of supply to demand for high-quality enlistees could fall short of its predrawdown levels. Furthermore, survey data show a drop in the rate at which potential high-quality recruits discuss military service with key "influencers" (such as family and friends) and fewer contacts between recruiters and high school students (perhaps due to cuts in numbers of recruiters, their reduced presence in high schools, or a shift in focus from current students to graduates). Taken together, these results suggest future difficulties in meeting accession goals, which should be countered by increases in recruiting resources such as advertising, educational benefits, and recruiters.

Author   Ozkaptan, Halim, William Sanders, and Robert F. Holz

Title   *Receptiveness of Army Families in USAREUR to Incentives for Extensions*

Report#   ADA174821

PubDate   1984

PubData   Alexandria, VA: U.S. Army Research Institute for the Behavioral and Social Sciences

Subject   retention, bonuses, job satisfaction, Army, military, personnel flow, development

Abstract

This research assesses the relative effects of a range of financial and other incentives on the willingness of Army families to extend tours in U.S. Army Europe (USAREUR) for 12 additional months. It also touches on some of the underlying reasons why families choose not to extend. Surveys were administered to more than 1,000 Army families, including both the service members and their spouses. The data obtained permit Army planners to determine the most cost-effective incentives for a desired extension rate. Data for enlistees, noncommissioned officers, and officers are presented separately.

Author   Palomba, C. A.

Title   *The Feasibility of Longer Enlistment Contracts*

Report#   ADA155979

PubDate   1984

PubData   Alexandria, VA: Center for Naval Analyses

Subject   military, accession, retention, cost

Abstract

This report determines that longer enlistment contracts are feasible for the Marine Corps. In addition to reviewing the literature, we examined data concerning the effect of contract length on accessions and on attrition. We estimated that requiring an additional year of enlistment is equivalent to an 8 percent pay reduction. We also found that attrition in the Marine Corps is not significantly affected by contract length. Using these findings, we determined that the cost per useful service year for five- and six-year enlistments is generally lower than for four-year enlistments.

Author   Patten, Thomas H.

Title   "How to Reform the Military Retirement System"

Report#

PubDate   1986

PubData   *Compensation and Benefits Review*, Vol. 18, No. 2, 1986, pp. 29–40

Subject   military, retirement, transitioning, costs, vesting, personnel flow

Abstract

The costs and structures of the U.S. military retirement system are reviewed. The retirement systems of Canada, the Federal Republic of Germany, the United Kingdom, and three U.S. private firms are also examined. The author makes recommendations for changes that would reduce costs, increase management flexibility, ease transitions from military life, and maintain trust by adequately serving the needs of military personnel. Recommendations are to retain 20-year vesting, delay benefits to age 55 with some transition help for younger retirees, add some partial benefits for individuals leaving between 10 and 20 years, include a grandfather clause, and continue a noncontributory system.

Author   Polich, J. M., James N. Dertouzos, and S. James Press

Title   *The Enlistment Bonus Experiment.*

Report#   RAND/R-3353-FMP

PubDate   1986

PubData   Santa Monica, CA : RAND

Subject   accession, recruiting, attrition, retention, high quality, personnel flow, bonuses, military, Army

Abstract

One of the principal challenges for defense managers in recent years has been to attract military recruits within a reasonable level of recruiting expenditure. This report describes the results of a nationwide experiment designed to provide new data on a key enlistment incentive: the cash enlistment bonus, which is paid to qualified recruits entering critical occupational specialties. The report documents the experiment, explains the analysis of its results, and assesses the effects of enlistment bonuses on the Army recruiting process. It addresses three principal effects of the bonus program: (1) attracting "high-quality" recruits into the Army; (2) encouraging enlistments in hard-to-fill critical specialties; and (3) influencing recruits to sign contracts for longer terms of service. The experimental results show that bonuses have substantial effects on recruiting and are a very flexible policy tool, making them a useful option for management of enlistment flows and for overcoming personnel shortages in critical skills.

Author   Price, Craig J.

Title   *An Application of the Job Characteristics Model to Enlisted Strategic Air Command Missile Maintenance Career Fields*

Report#   ADA161685

PubDate   1985

PubData   Wright-Patterson Air Force Base, OH: Air Force Institute of Technology

Subject   military, Air Force, development, job satisfaction, performance

Abstract

This thesis investigates the job attitudes of enlisted missile maintenance technicians performing duty at each of the Strategic Air Command's six Minuteman missile wings. The overall objective of the research was to determine whether a job enrichment program might hold potential for enhancing both the quality of work life and the individuals' work motivation. The methodology consisted of measuring levels of worker satisfaction with several dimensions of the work and work environment. The instrument used to collect sample data was the "Job Diagnostics Survey." Results show that a comprehensive job enrichment program is not warranted; however, attention should be given to contextual job factors.

Author   Quenette, Mary A.

Title   *Navy-Wide Personnel Survey 1993: Statistical Table for Enlisted Personnel*

Report#   ADA279044

PubDate   1994

PubData   San Diego, CA: Navy Personnel Research and Development Center

Subject   job assignment, morale, health, job satisfaction, Navy, military, employee involvement

Abstract

The fourth annual Navy-Wide Personnel Survey was mailed in September 1993 to a random sample of 17,902 active duty enlisted personnel and officers. Completed questionnaires were accepted through mid-December 1993. The adjusted return rate was 44 percent. Survey topics included detailing and assignment process, quality-of-life programs, leadership training, organizational climate, and health issues. Responses of enlisted personnel are broken out by pay grade and other important demographic variables.

Author   Quenette, Mary A.

Title   *Navy-Wide Personnel Survey (NPS) 1991: Management Report of Findings*

Report#   ADA255817

PubDate   1992

PubData   San Diego, CA: Navy Personnel Research and Development Center

Subject   retention, job satisfaction, child care, Navy, military, employee involvement, development

Abstract

The second annual NPS was mailed to 23,821 randomly sampled active duty enlisted personnel and officers in December 1991. Completed questionnaires were accepted through mid-February 1992. An adjusted return rate of 57 percent was obtained. Survey topics included rotation/permanent-change-of-station moves, recruiting duty, pay and benefits, education and leadership programs, quality-of-life programs, organizational climate, and AIDS education. This volume contains highlights of the results for enlisted personnel and officers.

Author   Quester, Aline, and Martha S. Murray

Title   *Attrition from Navy Enlistment Contracts*

Report#   ADA168117

PubDate   1986

PubData   Alexandria, VA: Center for Naval Analyses

Subject   recruiting, accession, DEP, attrition, background characteristics, military

Abstract

This research memorandum reports on the construction of an individual-based data set for Navy enlistment contracts for 1983 and 1984. It discusses the problems associated with these data and formalizes a contract attrition model, which is then estimated in a logistic framework. Findings show that demographic variables other than gender have only small effects on contract attrition. This is very different from studies of active duty attrition where these variables are shown to have much larger effects on attrition.

Author	Quester, Aline, and Sarah Jeffries
Title	*The History and Effectiveness of the Enlistment Bonus Program for Procuring Nuclear-Field Personnel*
Report#	ADA170881
PubDate	1985
PubData	Alexandria, VA: Center for Naval Analyses
Subject	bonuses, recruiting, accessions, compensation, skill needs, personnel flows, Navy, military

Abstract

Enlistment bonuses are monetary incentives promised to potential recruits to induce them to sign contracts to join the Navy. The bonuses, paid upon successful completion of class A schools, have been awarded in military skill areas characterized by inadequate volunteer levels. This memorandum describes how the Navy has used enlistment bonuses. The efficacy of the enlistment bonus program is clearly established by its significance for three measures of recruit procurement and across the specifications of the economic climate. Increasing the bonus by $1,000 adds about 30 new obligors to the DEP, increases total shipments by about 18 recruits, and reduces the number of nuclear-field recruits the Navy will be short by about 18.

Author	Quester, Aline O., and Adebayo M. Adedeji
Title	*Reenlisting in the Marine Corps: The Impact of Bonuses, Grade, and Dependency Status*
Report#	ADA246494
PubDate	1991
PubData	Alexandria, VA: Center for Naval Analyses
Subject	military, development , accession, promotion, SRB, bonuses

Abstract

First-term reenlistment decisions for recommended and eligible Marines in FY 1980 through FY 1990 are analyzed in this research memorandum. Particular attention is given to the retention effects of selective reenlistment bonuses on Marines in different Armed Force Qualification Test score categories. Additionally, reenlistment behavior for Marines of different marital statuses, grade, and length of initial enlistment contracts is analyzed. In the recent past, there have been substantial changes in the characteristics of enlisted Marines, as well as changes in Marine Corps personnel policy. First, enlisted Marines today are both smarter and better educated than they were in the earlier years of the 1980s. Second, although the percentage of recruits who enter the Marine Corps married or with dependents has remained virtually unchanged over time, the Marine Corps has experienced substantial increases in the marriage and dependency rate for enlisted personnel. Third, first-term enlistment contracts have been lengthened so that Marines now average more years of service at the first reenlistment point. Finally, there has been an increase in both time in service (TIS) and time in grade (TIG) for promotions to corporal and sergeant. The effect of these changes on the reenlistment decisions of first-term enlisted personnel (Zone A decisions) is the subject of this research memorandum.

Author   Quester, Aline O., and Greg W. Steadman

Title   *Enlisted Women in the Marine Corps: First-Term
            Attrition and Long-Term Retention*

Report#   ADA230764

PubDate   1990

PubData   Alexandria, VA: Center for Naval Analyses

Subject   military, Marines, attrition, accession, gender, retention,
            background characteristics

Abstract

This research memorandum provides an overview of gender differences in the continuation rates for enlisted Marines. It also contains a detailed analysis of first-term attrition for female recruits with four-year obligations accessed in FY1981 through FY1985. In this analysis, female first-term attrition probabilities are modeled as a function of background characteristics at entry into the Marine Corps. Over the first term of service, female attrition rates are about 1.5 times male attrition rates. Over the long term, female enlisted Marines have higher retention rates than male Marines. After 114 months of service 12.0 percent of female and 8.7 percent of male accessions were still in service.

Author   Quester, Aline O., and Janice A. Olson

Title   *Performance of Non-Prior-Service Navy Recruits: 1978–1986*

Report#   ADA203424

PubDate   1988

PubData   Alexandria, VA: Center for Naval Analyses

Subject   military, Navy, retention, accession, screening, background characteristics, attrition, performance

Abstract

This research memorandum describes measures of recruit success in the Navy using five performance indicators: desertion, demotion, first-term attrition, promotion, and retention. The CNA data base used here includes all nonprior service accessions between 1978 and December 1986. Differences in education swamp all other demographic characteristics in explaining differences in performance on all five indicators. Also, in terms of performance, a GED is not equivalent to a high school diploma.

Author   Rakoff, Stuart H., J. D. Griffith, and G. A. Zarkin

Title   *Models of Soldier Retention*

Report#   ADA282670

PubDate   1994

PubData   Alexandria, VA: U.S. Army Research Institute for the Behavioral and Social Sciences

Subject　　military, Army, retention, development, promotion, transitioning

Abstract

This report examines the effect of Army work, community, family, and individual factors on young soldiers' reenlistment intentions and outcomes. The data used in the report were collected in an Army-wide survey of a probability sample of soldiers and spouses conducted in 1989–1990. Retention intentions were examined for male soldiers in the ranks of private through staff sergeant for whom supervisor ratings of performance were available in the survey (n=5,299). Actual retention outcomes are examined for soldiers whose service obligation ended between the time of the survey and June 1990 (n=1,537). High retention intentions for high-performing soldiers were associated with favorable Army opportunities and a favorable climate for family life. Retention intentions were strong predictors of actual retention behavior.

Author　　Rearden, Anne-Marie

Title　　*A Multivariate Analysis of Reenlistment Intentions as a Predictor of Reenlistment Behavior*

Report#　　ADA206144

PubDate　　1988

PubData　　Monterey, CA: Naval Postgraduate School

Subject　　military, Navy, retention, transitioning, development

Abstract

The purpose of this thesis is to determine whether reenlistment intentions can help to predict actual reenlistment behavior. The sample consists of 6,328 Navy male enlisted members who are within one year of the reenlistment decision. The results show that the most powerful predictor of reenlistment behavior is the reenlistment intentions variable. The results also show that reenlistment behavior is influenced by race, age, pay grade, marital status, enlistment period, and level of satisfaction with the military in general.

Author   Robbert, A. L., Brent Keltner, Ken Reynolds, Mark Spranca, and Beth Benjamin

Title   *Differentiation in Military Human Resource Management*

Report#   RAND/MR-838-OSD

PubDate   1997

PubData   Santa Monica, CA: RAND

Subject   compensation, promotion, performance, teams, lateral entry, cost, military culture, training, gainsharing, retention, military, development

Abstract

The objectives of this research were to examine differentiated processes and outcomes in military HRM, to evaluate the suitability of these differentiations within the context of broad military organizational objectives, to identify and evaluate potentially enhancing alternatives, and to examine approaches for testing or implementing the more promising alternatives. The authors conclude that current military HRM works well in motivating many desired behaviors. Nonetheless, some functional communities today might benefit from greater differentiation while future environments may demand even more differentiation. Potential for differentiation is seen as being higher outside of core combat activities because shifts from hierarchical channeling of authority—a necessary element in many forms of differentiation—are not compatible with other exigencies in a combat environment. Cautions offered include a need to avoid undermining a powerful and well-focused system of intrinsic rewards and a need to support increased differentiation with better alternative accountability mechanisms.

Author   Rodney, David

Title   *Mathematical Relationships Between Pay Grade Structure, Longevity, and Promotion Policy*

Report#   ADA212150

PubDate   1988

PubDate   1988

PubData   Alexandria, VA: Center for Naval Analyses

Subject   military, promotion, development, transitioning, compensation

Abstract

Navy manpower requirements describe the number of personnel required to fulfill Navy missions. These manpower requirements are described by a variety of terms, including skill, pay grade, and length of service. Navy personnel managers institute numerous policies to attain and then to maintain the required number of personnel. In particular, compensation levels are systematically varied to obtain desired continuation rates, and promotion policy may be altered to attain desired numbers of personnel in each pay grade and longevity profiles within pay grades. It was concluded that stringent up-or-out policies are precisely the situations in which desired force structure becomes unobtainable. This is because the ability to successfully manage personnel can be shown to be directly related to the amount of permissible overlap in longevity between different pay grades. For example, suppose E-4 personnel are not allowed to have more than eight years of service, and promotion opportunity is such that only a few E-5 personnel have less than eight years of service.

Author   Rosen, Sherwin

Title   "The Military as an Internal Labor Market: Some Allocation, Productivity, and Incentive Problems"

Report#

PubDate   1992

PubData   *Social Science Quarterly*, Vol. 73, No. 2, 1992, pp. 227–237

Subject   compensation, promotion, pay table, performance, personnel flow, military, ability sorting, development

Abstract

Recent results from the economics of organizations are applied to military manpower. The military internal labor market must identify talented persons, provide adequate training, and efficiently assign them to positions. Promotions and wage policy must maintain proper incentives for retention and motivation. These organizational needs are not efficiently met in terms of the legacy of military pay compression and paternalistic personnel policies inherited from the conscription era.

Author	Rosenberg, Florence R., and Jocelyn S. Vuozzo
Title of	*Thematic Analysis of Spouse Comments: Annual Survey Army Families, 1987*
Report#	ADA212035
PubDate	1989
PubData	Washington, DC: Walter Reed Army Institute of Research
Subject	military, Army, family, development

Abstract

This research consists of an analysis of comments made by Army spouses in the 1987 ASAF survey. About 40 percent of the 12,000 survey respondents volunteered opinions, criticisms, suggestions, etc., on a wide variety of Army-related topics. Comments were coded under 25 major subject areas, many including sub-topics.

Author	Rosenfeld, Paul, Amy L. Culbertson, and Carol E. Newell
Title	*Assessment of Equal Opportunity Climate: Results of the 1991 Navy-Wide Survey*
Report#	ADA302861
PubDate	1995
PubData	San Diego, CA: Navy Personnel Research and Development Center
Subject	gender, equal opportunity, Navy, military

Abstract

This report describes the results of the 1991 Navy Equal Opportunity/Sexual Harassment (NEOSH) Survey. The survey was administered to a random sample of active duty Navy enlisted personnel and officers stratified on racial/ethnic group and gender. The first NEOSH Survey was conducted in 1989.   The magnitude of race/ethnic and gender differences in perceptions were less in 1991 than in 1989.

Author   Ross, Robert M., Glenda Y. Nogami, and Newell K. Eaton

Title   *The Impact of Occupational Specialty and Soldier Gender on First Tour Enlisted Attrition*

Report#   ADA163502

PubDate   1984

PubData   Alexandria, VA: U.S. Army Research Institute for the Behavioral and Social Sciences

Subject   attrition, gender, accession, personnel flow, job assignment, Army, military

Abstract

The purpose of this research was to determine the effect of individual and organizational variables on first-tour soldier attrition. All female recruits and a 10 percent sample of noncombat male recruits entering the Army in FY76 were studied. Results showed major differences by gender, type of attrition, traditionality of MOS, education, and race. Females had a higher attrition rate overall. Also, females had higher attrition due to family-related reasons, and males had higher attrition due to adverse causes. Females in nontraditional MOSs had higher attrition rates. High school graduation was the single best predictor of first-term success.

Author   Royle, Marjorie H.

Title   *Factors Affecting Attrition Among Marine Corps Women*

Report#   ADA162537

PubDate   1985

PubData   San Diego, CA: Navy Personnel Research and Development Center

Subject   gender, attrition, recruiting, accessing, job satisfaction, military, Marine Corps, development

Abstract

A representative sample of Marine Corps women in their first enlist-ments and their supervisors were surveyed to identify factors in their backgrounds and experiences that might be related to attrition. The most important factors were supervisor and work group relation-ships, family and career orientation, and management of stress. Recruiting, training, and placement practices had a relatively small effect on attrition. To help decrease attrition among women, the Marine Corps should discourage the most traditional women from enlisting, help women develop coping skills, provide sex education, and improve work group climate as well as climate toward women as a whole in the USMC.

Author	Rydell, Peter C.
Title	*ALEC : A Model for Analyzing the Cost-Effectiveness of Air Force Enlisted Personnel Policies (Theory and Results)*
Report#	RAND/N-2629/1-AF
PubDate	1987
PubData	Santa Monica, CA : RAND
Subject	personnel flow, compensation, bonuses, accession, transitioning, retention, cost, training, Career-Job Reservation (CJR), military, Air Force, development

Abstract

The Aggregate Lifecycle Effectiveness and Cost (ALEC) model en-ables managers of Air Force enlisted personnel to compare the cost effectiveness of alternative management actions for a part of the force selected for analysis. Example actions are limits on the num-bers of various types of enlistments, reenlistment bonuses designed to increase the number of persons making the Air Force a career, re-training programs that transfer personnel from one specialty to an-other, and the early-release program. This volume gives the theory and behavioral relationships used to build the model and gives cost-effectiveness results.

Author   Schaffer, Robert J.

Title   *Relating the Armed Services Vocational Aptitude Battery to Marine Job Performance*

Report#   ADA318946

PubDate   1996

PubData   Monterey, CA: Naval Postgraduate School

Subject   military, Marines, training, development, accessing, performance, screening

This thesis develops a method to reconfirm the relationship between an individual's Armed Forces Vocational Aptitude Battery scores and his performance at his initial course of instruction in the Marine Corps. Once the ASVAB is shown to correctly predict success at Marine Corps courses, the thesis concentrates on two statistical methods to explore the classification of youths into Marine jobs. These methods ultimately afford the Marine Corps an opportunity to use existing information to enhance the successful classification of young Marines into appropriate courses.

Author   Shanley, Michael G., John D. Winkler, and Paul Steinberg

Title   *Resources, Costs, and Efficiency of Training in the Total Army School System*

Report#   RAND/MR-844-A

PubDate   forthcoming

PubData   Santa Monica, CA: RAND

Subject   training, cost, Army, military, development

Abstract

This report analyzes the resource use and efficiency of the reserve components' new prototype school system in the southeast section of the United States (Region C). The assessment of outcomes in fiscal year 1995 (the execution year of the prototype) is based on data collected in both fiscal year 1994 (the baseline year) and 1995 in both Region C and E, a comparison region in the Midwest. The document also discusses ways to further improve resource use and efficiency in the future—primarily by more fully utilizing school system capacity. This report is part of a larger effort by RAND Arroyo to analyze the performance and efficiency of the RC school system.

Author	Sheposh, John P., Joyce S. Dutcher, and Carol A. Hayashida
Title	*Experimental Civilian Personnel Office Project (EXPO): Final Report for Appropriated Fund Sites*
Report#	ADA277447
PubDate	1994
PubData	San Diego, CA: Navy Personnel Research and Development Center
Subject	military, development, performance

Abstract

The purpose of this report is to document the evaluation of the Experimental Civilian Personnel Office (EXPO) project at appropriated fund (APF) sites. The aim of the EXPO project was to design and test initiatives that streamlined and simplified personnel management procedures and policies employed in the Department of Defense. Sixteen APF sites were evaluated: one U.S. Air Force, one U.S. Army, one U.S. Navy, and 13 Defense Logistics Agency sites. The initiatives signified a movement toward greater empowerment of nonpersonnelists (e.g., managers), greater efficiency, flexibility, and responsiveness. The evaluation covered a three-year test period. Data sources included attitudinal surveys, standardized on-site interviews of personnelists, managers, supervisors, and existing personnel data bases. The results indicate that EXPO had a positive effect.

Author	Smith, Keith E.
Title	*Cost Comparison of Technical School Versus Unit Training Methods for Direct-Duty Airmen in Civil Engineering Air Force Specialty Codes*
Report#	ADA174565

PubDate   1986

PubData   Wright-Patterson Air Force Base, OH: Air Force Institute of Technology

Subject   cost, training, development, Air Force, military

Abstract

This study compares the productivity of civil engineering airmen who were either trained in technical school or sent to direct-duty to be trained by active squadrons. The results of the research show that technical training school is more cost-effective. Cost savings range from $18,784 to $43,508 per recruit for the seven specialties examined. Cost savings are related to lost time and production within active units for both trainees and the trainers. Savings are all found in the first year of service with subsequent years showing no difference between the airmen trained by the two methods.

Author .   Stafford, Frank P.

Title   "Partial Careers: Civilian Earnings and the Optimal Duration of an Army Career"

Report#

PubDate   1991

PubData   Alexandria, VA: U.S. Army Research Institute for the Behavioral and Social Sciences (in *Military Compensation and Personnel Retention*, edited by C. L. Gilroy, D. K. Horne, and D. A. Smith)

Subject   military, Army, compensation, transitioning, develop ment

Abstract

This paper provides a conceptual framework for studying military careers and uses a special merger of detailed Army records with Internal Revenue Service and Social Security Administration earnings records to examine the post-service labor market experience of U.S. Army "alumni." It is found that military occupational specialty, years of military and civilian experience, speed of promotion, race, and the state unemployment rate have an influence on earnings.

The results imply that the optimal Army pay policy needs to be targeted to military occupational specialty.

Author   Steel, Robert P.

Title   "Labor Market Dimensions as Predictors of the Reenlistment Decisions of Military Personnel"

Report#

PubDate   1996

PubData   *Journal of Applied Psychology*, Vol. 81, No. 2, pp. 421–428.

Subject   military, reenlistment, retention, development, personnel flow

Abstract

Labor market variables (e.g., unemployment statistics) and perceptual measures of employment opportunities (e.g., perceived occupational demand) were used to predict the reenlistment decisions of 402 U.S. Air Force enlistees. With logistic regression analysis, three significant predictors of reenlistment were isolated. A combination of perceptual and objective job availability measures provided the best predictors of the reenlistment criterion.

Author   Stephenson, Stanley P., Margaret E. Mitchell, L. L. Beik, D. R. Ellison, S. D. Fitch, and D. A. Macpherson

Title   *Factors Which Affect Motivations for Staying In, Leaving, and Reentering the United States Navy*

Report#

PubDate   1984

PubData   Brussels, Belgium: Symposium Papers from the Second Symposium on Motivation and Morale in NATO Forces

Subject   accession, recruiting, attrition, retention, personnel flow, Navy, military, background characteristics

Abstract

This paper presents results from an empirical study of enlisted men in the United States Navy for the period 1973 to 1982. It focuses on issues related to individual decisions to stay in, leave, or reenter the Navy. Results show how a range of background characteristics and job categories affect these three decisions.

Author	Stolzenberg, Ross M., and John D. Winkler
Title	*Voluntary Terminations from Military Service*
Report#	RAND/R-3211-MIL
PubDate	1983
PubData	Santa Monica, CA: RAND
Subject	recruiting, performance, personnel flow, accession, job assignment, costs, attrition, military, compensation, recruit screening

Abstract

This report attempts to integrate and critique the differing theoretical perspectives, methodologies, populations analyzed, and findings of several studies of why personnel leave the services. The core of the report's theoretical framework is Thibaut and Kelley's research on how people evaluate their membership in groups. It considers the influence of the following factors on voluntary terminations from military service: compensation; job security and dispute resolution procedures; amenities, conveniences, psychological rewards, and working conditions; and individual differences in pre-service attributes and demographic characteristics. The analysis suggests that terminations can be reduced by making personnel aware of the true value of their compensation, and using lump sum payments more extensively. Nonpecuniary factors also affect terminations. Findings indicate that mechanisms for resolving disputes improve retention; realistic portrayal of service life to new recruits fosters their retention; and voluntary termination is more likely if the recruit has a history of antisocial behavior, lacks a high school diploma, has a spouse and dependent children, or is under 18 years of age. The authors conclude that new data and new analyses are needed to make significant advances in knowledge about the causes of voluntary termination from military service.

Author   Thomlison, Cynthia Ann

Title   *Fundamental Applied Skills Training (FAST) Program Measures of Effectiveness*

Report#   ADA310383

PubDate   1996

PubData   Monterey, CA: Naval Postgraduate School

Subject   training, accession, attrition, personnel flow, development, military, Navy

Abstract

This thesis attempts to measure the effectiveness of Fundamental Applied Skills Training (FAST), a program designed to help selected Navy recruits succeed in Basic Military Training (BMT) by improving their literacy skills. The study analyzes whether completion of FAST is related to subsequent completion of BMT, completion of first term of service, and advancement toward higher rank (E-4) for recruits who entered the Navy in Fiscal Years 1992 and 1993. Study results suggest that participation in FAST is related to an increased probability of completing BMT and generally higher success chances in the Navy during the first term of service.

Author   Thompson, Theodore J., Iosif A. Krass, and Timothy T. Liang

Title   *Quantifying the Impact of the Permanent Change of Station (PCS) Budget on Navy Enlisted Personnel Unit Readiness*

Report#   ADA237162

PubDate   1991

PubData   San Diego, CA: Navy Personnel Research and Development Center

Subject   development, performance, cost, job assignment, military

Abstract

The Navy spends over $600 million annually to move its active duty personnel. This report describes a model that relates permanent change of station (PCS) moves to personnel readiness. To support this model, a readiness measure, which can distinguish among small differences in manning levels, was developed. Using a hypothetical PCS move plan for a five-month time horizon we show the effect of a reduced moving budget. The current move plan is for 17,300 moves whereas the reduced budget plan calls for 14,495 moves. The reduction in moves translates to a drop in readiness from 1.85 to 2.30.

Author   Toomepuu, Juri

Title   *Costs and Benefits of Quality Soldiers: A Critical Review of the CBO Report, Quality Soldiers: Costs of Manning the Active Army*

Report#   ADA173223

PubDate   1986

PubData   Fort Sheridan, IL: U.S. Army Recruiting Command

Subject   cost, accession, recruiting, quality, performance, personnel flow, Army, military, readiness

Abstract

This research note reviews the CBO report and its cost-effectiveness and productivity measures, the assumptions about productivity and the effects of bonuses and military pay on Army enlistments, and the conclusions drawn. The critique finds the cost-effectiveness and productivity measures used, as well as elasticities of resources, inappropriate, and provides alternatives that lead to different conclusions. Although recruitment of high-quality youth is costly, quality manpower is cost-effective from a larger perspective. The critique also points out that many hidden and incommensurable costs were not addressed by the CBO, and concludes that the program proposed by the CBO would substantially lower the Army's combat- and cost-effectiveness.

Author   United States Army Sergeants Major Academy

Title   *The Duties, Responsibilities, and Authority of NCO's*

Report#   ADF250227

PubDate   1981

PubData   Biggs Army Air Field, TX: United States Army Sergeants Major Academy

Subject   work systems, Army, military, development,

Abstract

This manual was prepared to further develop and refine the doctrine of the functional relationship which must exist between officers and NCOs at all levels within the Army. Included is the essential doctrine necessary for commissioned and noncommissioned officers to begin the process of clarifying their duties, responsibilities, and authority.

Author   Veit, C. T.

Title   *Effects of Apache Helicopter Crew and Unit Training on Combat Mission Effectiveness*

Report#   RAND/P-7590

PubDate   1989

PubData   Santa Monica, CA: RAND

Subject   training, performance, military, development

Abstract

Using algebraic modeling and subjective transfer function testing methods, the research presented in this paper examines the effects of training and technology on crew proficiency and combat mission effectiveness for the Apache's close-support day mission. Preliminary findings suggest that increasing training regimens can produce substantial increases in combat effectiveness, similar in magnitude to using the LHX (Light Helicopter Experimental) as the Apache's scout, or a capability that locates targets and threats more accurately. The data suggest guidelines for changing training programs.

Author	Waite, Linda J., and Sue E. Berryman
Title	*Women in Nontraditional Occupations: Choice and Turnover*
Report#	RAND/R-3106-FF
PubDate	1985
PubData	Santa Monica, CA: RAND
Subject	recruit screening, gender, military, civilian, attrition, retention, accession

Abstract

This report uses data from the National Longitudinal Survey of Youth Labor Market Behavior to test a series of hypotheses about characteristics of individuals and their families that influence their occupational preferences and their turnover in the military and in civilian jobs. The study's findings have three important policy implications: (1) Women enlistees have much lower exit rates from the armed forces than their counterparts in civilian jobs; (2) job traditionality does not affect turnover for women in civilian jobs (for a variety of definitions of the traditionality variable and for several alternative specifications of the civilian turnover model); and (3) for women in the military there is no effect of being in a traditionally female or a traditionally male occupation on turnover.

Author	Walker, Warren E.
Title	*Design and Development of an Enlisted Force Management System for the Air Force*
Report#	RAND/R-3600-AF
PubDate	1991
PubData	Santa Monica, CA: RAND
Subject	personnel flow, bonuses, retention, attrition, retirement, military, Air Force, compensation, development, transitioning

Abstract

This report provides an overview of the design of the Air Force's Enlisted Force Management System (EFMS) and how that design is being implemented. The EFMS is a decision support system whose purpose is to improve the effectiveness and efficiency of the members of the Air Staff who are engaged in managing the enlisted force. This report describes the concepts underlying the EFMS and provides brief introductions to the system's models and databases. It also provides references to other documents that contain more detailed information on specific aspects of the system.

Author	Ward, Michael P., and Hong W. Tan
Title	*The Retention of High-Quality Personnel in the U.S. Armed Forces*
Report#	RAND/R-3117-MIL
PubDate	1985
PubData	Santa Monica, CA: RAND
Subject	retention, military, background characteristics, quality, recruit screening, accession, attrition, personnel flow, promotion, development

Abstract

This study addresses the question, Does the military retain the best of its first-term recruits? Using data from the 1974 Entry Cohort File developed by the Defense Manpower Data Center, the authors generate an index of job performance that combines entry-level attributes of recruits—Armed Forces Qualification Test scores and level of education—with first-term promotion histories. This "quality index" is used to assess the relative importance of these characteristics and other unobserved "ability factors" for evaluating the military's success in retaining high-quality enlisted personnel. The authors find that the military is, in general, successful in retaining high-quality enlisted personnel. Those lost through attrition have the lowest quality. Those who separate at the end of their commitment have about the same quality as those entering the military. The study is a first step toward answering the important policy question of how the military can attract and retain high-quality recruits, and how reenlistment standards should be designed.

Author	Warner, John T., and Gary Solon
Title	"First-Term Attrition and Reenlistment in the U.S. Army"
Report#	
PubDate	1991
PubData	Alexandria, VA: U.S. Army Research Institute for the Behavioral and Social Sciences (in *Military Compensation and Personnel Retention*, edited by C. L. Gilroy, D. K. Horne, and D. A. Smith)
Subject	military, Army, accession, attrition, retention, costs, screening

Abstract

These data present an analysis of new data on first-term attrition and the first reenlistment decision. The analysis uses data on men who enlisted in Infantry MOSs between 1974 and 1983. Probit, logit, and proportional hazard models are used to analyze first-term attrition and reenlistment. Findings show that high school graduates are much more likely to survive the first term, but less likely to reenlist, minorities are more likely to survive and to reenlist, and reenlistment decisions are responsive to the pecuniary attractiveness of Army versus civilian employment.

Author	Warner, John T., and Matthew S. Golderg
Title	"The Influence of Non-Pecuniary Factors on Labor Supply: The Case of Navy Enlisted Personnel"
Report#	
PubDate	1984
PubData	*Review of Economics and Statistics*, Vol. 66, 1984, pp. 26–35
Subject	retention, duty assignment, personnel flow, compensation, bonuses, military, Navy, development

Abstract

This paper expresses the elasticity of labor supply to an occupation in terms of the variables and covariances of the distributions of non-pecuniary factors across the alternative occupations. We then apply our results to an analysis of reenlistment rates among Navy enlisted personnel. We argue that those Navy skills experiencing a higher incidence of sea duty will have lower pay elasticities. This prediction is confirmed empirically using probit analysis.

Author	Wilcove, Gerry L., R. L. Burch, A. M. Conroy, and R. A. Bruce
Title	*Officer Career Development: A Review of the Civilian and Military Research Literature on Turnover and Retention*
Report#	ADA241363
PubDate	1991
PubData	San Diego, CA: Navy Personnel Research and Development Center
Subject	military, civilian, management, accession, retention, attrition

Abstract

Reviews were conducted of both civilian and military research literature on turnover in preparation for a predictive study involving aviation warfare officers and actual retention behavior. In the present report, results from the two literature reviews are presented and compared.   An annotated bibliography, which has been computerized, is also presented, and is available on disk upon request.

Author	Wild, William G., and Bruce R. Orvis
Title	*Design of Field-Based Crosstraining Programs and Implications for Readiness*
Report#	RAND/R-4242-A
PubDate	1993

PubData   Santa Monica, CA: RAND

Subject   training, cost, Army, military, development, personnel flow

Abstract

As part of a broad effort to reduce defense expenditures, the Army is exploring a number of new approaches to training individual soldiers. Prominent among these approaches is "field-based crosstraining" (FBCT), which involves combining two or more occupational specialties and shifting initial skill training from Army schools to on-the-job training in field units. This report describes a method for analyzing the features, advantages, and disadvantages of field-based crosstraining programs in the Army. Focusing on the specific case of helicopter maintenance, the report analyzes data from field units and recommends alternative field-based cross-training strategies for the Army. An assessment of the Army's Apprentice Mechanic Initiative (AMI) for helicopter maintenance is included in the analysis.

Author   Winkler, John D., Stephen J. Kirin, and John S. Uebersax

Title   *Linking Future Training Concepts to Army Individual Training Programs*

Report#   RAND/R-4228-A

PubDate   1992

PubData   Santa Monica, CA: RAND

Subject   training, costs, military, development,

Abstract

This report presents the results of research seeking to link new Army training concepts for changing institutional training programs to specific occupations and courses. It analyzes, across a range of occupations, alternative training approaches that may be more affordable and flexible than current techniques for individual skill training. The report examines training-related characteristics of Army occupations and identifies general training-related dimensions that characterize Army entry-level enlisted military occupational specialties. The authors find that the principal training-related dimensions include ability requirements, dominant task characteristics (procedural or verbal), similarity to civilian occupations, and resource intensity. The dimensions can be linked to new training concepts under consideration by the Army (i.e., distributed training; use of training aids, devices, simulators, and simulations; use of civilian training sources). The authors find these results useful in suggesting MOS in which particular training concepts and strategies may prove most feasible and cost-effective.

Author	Winkler, John, D., and Paul Steinberg
Title	*Restructuring and Consolidation of Military Education and Training*
Report#	RAND/MR-850-A/RC
PubDate	1997
PubData	Santa Monica, CA: RAND
Subject	training, development, cost, performance, civilian substitution, military

Abstract

This report draws on the findings of numerous RAND studies of military education and training conducted over the last two decades and identifies tools and provides insights for making training more efficient and affordable. The findings are drawn largely from studies that addressed individual military education and training, which provides soldiers with job-specific skills and knowledge needed to perform their functions as members of military organizations. Also addressed are the implications of this research for other types of training (e.g., collective training in units) and for functions related to individual training that are customarily not analyzed (e.g., training development and support).

Author	Winkler, John, D., Judith C. Fernandez, and J. Michael Polich
Title	*Effect of Aptitude on the Performance of Army Communications Operators*
Report#	RAND/R-4143-A

PubDate   1992

PubData   Santa Monica, CA: RAND

Subject   accession, job assignment, performance, quality, Army, military, AFQT, development

Abstract

This report examines duty tasks performed by military occupational specialty 31M, Multichannel Communications Equipment Operator, whose members operate communications systems providing division- and corps-level command and control. The intent was to develop quantitative analyses based on objective measurement of soldier and unit performance aimed at improving the Army's ability to set appropriate performance standards and to develop quantitative estimates of the link between personnel aptitude and Army operational performance. The study finds that the Armed Forces Qualification Test score has a direct, consistent effect on the ability of communications personnel to provide effective battlefield communications to Army units. The evidence suggests that AFQT scores have a sizable effect on group performance. Groups that are on average "smarter" outperform other groups. The study concludes that a lowering of accession standards will substantially reduce the probability of operator success in operating and troubleshooting communications systems.

Author   Yardley, Roland James

Title   *An Analysis of the Effect of ASVAB Waivers on A-School Academic Attrition*

Report#   ADA241788

PubDate   1990

PubData   Monterey, CA: Naval Postgraduate School

Subject   military, Navy, training, screening, performance, quality, development

Abstract

The purpose of this thesis was to analyze the effect of ASVAB waivers on A-School academic attrition. This was accomplished utilizing extracts of the Enlisted Training Tracking File and the Navy Enlisted Classification Tracking File. The data base was explored by conducting an analysis of those individuals who did not have the prerequisite ASVAB score, and then comparing their A-School academic performance with those who had attained the prerequisite. Significant differences in various measures of performance were found between the two groups, including disenrollment and academic attrition.

Author	Zimmerman, Dona C., Ray A. Zimmerman, and William H. King
Title	*Development and Validation of Preenlistment Screening Composites for Army Enlisted Personnel*
Report#	ADA165236
PubDate	1985
PubData	Fort Sheridan, IL: Program Analysis and Evaluation Directorate
Subject	recruiting, accession, attrition, recruit screening, background characteristics, military, Army

Abstract

This report describes the results of analysis employed to develop and compare male and female preenlistment suitability screens. The population studied consisted of nearly 300,000 male and 51,000 female nonprior service recruits who enlisted from 1979 to 1982. Predictor variables considered in the development of the screening composites included educational level, AFQT category, age, race, term of enlistment, and other entry factors.

Author	Zimmerman, Ray A., and Dona C. Zimmerman
Title	*Recruitment of College-Bound Youth Through Use of the ACT Assessment File*
Report#	ADA165582
PubDate	1985
PubData	Fort Sheridan, IL: Program Analysis and Evaluation Directorate

Subject   recruiting, accession, quality, skill needs, military, background characteristics

Abstract

The purpose of this study was to examine the effectiveness of using the American College Testing (ACT) assessment records in recruiting "college bound" young people to fill highly specialized enlisted positions. Exploratory studies were conducted to examine the utility of telephone interviews and mail campaigns in stimulating interest among high school students in foreign language training at the Defense Language Institute. Both approaches were very effective in attracting individuals from the target segment of the college-bound market. The telephone interviews were more effective, but also more costly. Results suggest further study in using this approach to fill specialized positions.

Author   Zimmerman, Ray C., Dona C. Zimmerman, and Mary Ellen Lathrop

Title   *Study of Factors Related to Army Delayed-Entry Program Attrition*

Report#   ADA166894

PubDate   1985

PubData   Fort Sheridan, IL: Program Analysis and Evaluation Directorate

Subject   accession, recruiting, attrition, DEP, military, Army

Abstract

The purpose of this study was to examine the relative influence of personal and situational factors on Delayed Entry Program accession/attrition decisions. In addition to demographic characteristics, this study focused on variables such as experiences during the recruitment process and valued outcomes the recruit expected to obtain from military service. Telephone interviews of 1,000 individuals participating in the DEP during FY1984 were conducted to gather information pertaining to individuals' valued outcomes, experiences in the recruiting process, perceptions of the job market conditions, participation in DEP activities, etc. The findings indicated that satisfaction with the occupational assignment was an important factor in accession/attrition decisions. Also important were the experiences of recruits during their tenure in the DEP. Implications for effective DEP management and pre-accession socialization are discussed.

Eitelberg, M. J., and S. L. Mehay, "Looking Ahead: The Shape of Things to Come," in M. J. Eitelberg and S. L. Mehay (eds.), *Marching Towards the 21st Century*, CT: Greenwood Press, 1994.

Keeney, R. L., *Value-Focused Thinking: A Path to Creative Decisionmaking*, Cambridge, MA: Harvard University Press, 1992.

Keeney, R. L., and C. W. Kirkwood, "Group Decision Making Using Social Welfare Functions," *Management Science*, Vol. 22, 1975, pp. 430–437.

Kirby, S. N., and H. J. Thie, *Enlisted Personnel Management: A Historical Perspective*, Santa Monica, CA: RAND, MR-755-OSD, 1996.

Kirkwood, C. W., *Multiobjective Decision Analysis*, Tempe, AZ: Arizona State University, 1995.

Kirkwood, C. W., and R. K. Sarin, "Preference Conditions for Multiattribute Value Functions," *Operations Research*, Vol. 28, 1980, pp. 225–232.

Levy, C. M., H. J. Thie, and J. Sollinger, "How Other Organizations Manage Their Workforces: Insights for Future U. S. Enlisted Force Management," unpublished RAND research.

Saaty, T. L., "How to Make a Decision: The Analytic Hierarchy Process," *Interfaces*, November–December 1994, pp. 19–43.

Thie, H., and R. A. Brown, *Future Career Management Systems for U.S. Military Officers*, Santa Monica, CA: RAND, MR-470-OSD, 1994.

Thie, Harry J., Margaret C. Harrell, Roger A. Brown, Clifford M. Graf II, Mark Berends, Claire M. Levy, and Jerry M. Sollinger, *A Future Officer Career Management System: An Objectives-Based Design*, Santa Monica, CA: RAND, MR-788-OSD, 2001.

White, J., "Commentary," in W. Bowman, R. Little, and G. T. Sicilia (eds.), *The All-Volunteer Force After a Decade: Retrospect and Prospect*, Washington: Pergamon-Brassey's, 1986.